Extraction Methods
in Organic Analysis

Sheffield Analytical Chemistry

Series Editor: R.N. Ibbett

A series which presents the current state of the art of chosen sectors of analytical chemistry. Written at professional and reference level, it is directed at analytical chemists, environmental scientists, food scientists, pharmaceutical scientists, earth scientists, petrochemists and polymer chemists. Each volume in the series provides an accessible source of information on the essential principles, instrumentation, methodology and applications of a particular analytical technique.

Titles in the Series:

Inductively Coupled Plasma Spectrometry and its Applications
Edited by S.J. Hill

Extraction Methods in Organic Analysis
Edited by A.J. Handley

Extraction Methods in Organic Analysis

Edited by

ALAN J. HANDLEY
LGC
Runcorn
Cheshire

CRC Press

First published 1999
Copyright © 1999 Sheffield Academic Press

Published by
Sheffield Academic Press Ltd
Mansion House, 19 Kingfield Road
Sheffield S11 9AS, England

ISBN 1-85075-941-3

Published in the U.S.A. and Canada (only) by
CRC Press LLC
2000 Corporate Blvd., N.W.
Boca Raton, FL 33431, U.S.A.
Orders from the U.S.A. and Canada (only) to CRC Press LLC

U.S.A. and Canada only:
ISBN 0-8493-9740-5

All rights reserved. No part of this publication may be reproduced, stored in a retrieval system or transmitted in any form or by any means, electronic, mechanical, photocopying or otherwise, without the prior permission of the copyright owner.

This book contains information obtained from authentic and highly regarded sources. Reprinted material is quoted with permission, and sources are indicated. Reasonable efforts have been made to publish reliable data and information, but the author and the publisher cannot assume responsibility for the validity of all materials or for the consequences of their use.

Trademark Notice: Product or corporate names may be trademarks or registered trademarks, and are used only for identification and explanation, without intent to infringe.

Printed on acid-free paper in Great Britain by
Bookcraft Ltd, Midsomer Norton, Bath

British Library Cataloguing-in-Publication Data:
A catalogue record for this book is available from the British Library

Library of Congress Cataloging-in-Publication Data:
Extraction methods in organic analysis / edited by Alan Handley.
    p.   cm. -- (Sheffield analytical chemistry)
  Includes bibliographical references and index.
  ISBN 0-8493-9740-5 (alk. paper)
  1. Extraction.   2. Organic compounds--Analysis.   I. Handley, Alan
(Alan John)   II. Series
QD63.E88E9   1998
547'.3--dc21
                                                                    98-4748
                                                                      CIP

Preface

Sample extraction plays an ever increasing role in modern analytical methodology. Similar to sampling, it provides one of the key steps prior to the final analytical determination, and, if carried out effectively, it assures the validity of the measurement. The explosion of interest in extraction and sample preparation methodologies in recent years has seen the introduction of many new techniques, fuelled by scientific and industrial demand for faster and simpler analysis. Much has been published on the topic in journals, and a considerable database of applications has evolved. However, there have been few books dedicated to extraction methods in organic analysis, especially the newer methods. The aim of this book is therefore to provide a complete, up-to-date and practical reference work for scientists and technologists involved in both sample preparation and analysis, whether in academic or industrial laboratories and irrespective of scientific discipline.

I have attempted to include all of the most important new developments in the area. Following a short introduction, chapter 2 sets the scene by describing the more classical methods used in liquid/liquid and liquid/solid extractions. Chapters 3 and 4 consider the applications and developments of Solid Phase Extraction (SPE), Solid Phase Micro Extraction (SPME), and Membrane Extraction with Sorbent Interface (MESI). Chapters 5, 6 and 7 describe the key alternative technologies for solid/liquid extraction: Supercritical Fluid Extraction (SFE), Pressurised Fluid Extraction (PFE) or Accelerated Solvent Extraction (ASE) and Microwave-Assisted Solvent Extraction (MAE). Each of these chapters is designed to provide an insight into the basic principles, the theory, the equipment, the strategies for method development, the applications and the future developments and directions. The final three chapters revisit the application areas in greater depth, examining the main areas for extraction technology: Biological and Pharmaceutical Analysis, Analysis in the fields of Polymers and Polymer Additives, and Environmental Monitoring.

To cover such a wide remit, authors have been sought from both the academic and industrial sectors, to bring the necessary expertise and balance to the volume. This, I feel, has helped to produce an informed and comprehensive review of the current status of organic extraction technology.

I would like to thank the authors for the time, energy and effort that went into the writing of each chapter, and for their excellent manuscripts, which required little or no editing.

A.J. Handley

Contributors

Dr Erik Baltussen	Laboratory of Instrumental Analysis, Eindhoven University of Technology, PO Box 513, 5600MB Eindhoven, The Netherlands
Dr Anthony A. Clifford	Express SFC Technology, 175 Woodhouse Lane, Leeds LS2 3AR, UK
Dr Frank David	Research Institute for Chromatography, Kennedypark, 20 B-8500 Kortrijk, Belgium
Dr Tom De Smaele	Department of Analytical Chemistry, University of Gent, Proeftuinstraat 86, B-9000 Gent, Belgium
Dr John R. Dean	Department of Chemical and Life Sciences, Ellison Building, University of Northumbria, Newcastle-upon-Tyne NE1 8ST, UK
Mr John L. Ezzell	Dionex Corporation, Suite A, 1515 West, 2200 South, Salt Lake City, Utah 84119, USA
Miss Lisa Fitzpatrick	Department of Chemical and Life Sciences, Ellison Building, University of Northumbria, Newcastle-upon-Tyne NE1 8ST, UK
Mr Alan J. Handley	LGC, The Heath, Runcorn WA7 4QD, UK
Miss Carolyn Heslop	Department of Chemical and Life Sciences, Ellison Building, University of Northumbria, Newcastle-upon-Tyne NE1 8ST, UK
Dr Alexis J. Holden	Centre for Toxicology, University of Central Lancashire, Preston PR1 2HE, UK
Dr Hans-Gerd Janssen	Laboratory of Instrumental Analysis, Eindhoven University of Technology, PO Box 513, 5600MB Eindhoven, The Netherlands
Dr Xianwen Lou	Laboratory of Instrumental Analysis, Eindhoven University of Technology, PO Box 513, 5600MB Eindhoven, The Netherlands

Dr Yuzhong Luo	Department of Chemistry, University of Waterloo, Ontario, ON N2L 3GI, Canada
Dr Robert D. McDowall	McDowall Consulting, Bromley, Kent, UK
Dr Steve Miller	Zeneca plc, Mereside, Aldersley Park, Macclesfield, SK10 4DG, UK
Professor Janusz Pawliszyn	Department of Chemistry, University of Waterloo, Ontario, ON N2L 3GI, Canada
Professor Pat Sandra	Department of Organic Chemistry, University of Gent, Krijgslaan 281-S4, B-9000 Gent, Belgium
Dr Derek Stevenson	School of Biological Sciences, University of Surrey, Guildford GU2 5XH, UK
Dr Harold J. Vandenburg	Department of Chemistry, University of Leeds, Leeds LS2 9JT, UK
Professor Ian D. Wilson	Zeneca plc, Mereside, Aldersley Park, Macclesfield, SK10 4DG, UK

Contents

1	**Introduction** A. J. HANDLEY	**1**
2	**Solvent and membrane extraction in organic analysis** A. J. HOLDEN	**5**

2.1 Introduction 5
2.2 Solvent extraction 5
    2.2.1 Basic principles 10
    2.2.2 Instrumentation and procedures 14
    2.2.3 Strengths and limitations 24
    2.2.4 General applications 24
2.3 Other methods involving solvent extraction and/or filtration procedures 25
    2.3.1 Basic principles 25
    2.3.2 Instrumentation and methods 30
    2.3.3 Strengths and limitations 31
    2.3.4 General areas of application 31
2.4 Membrane extraction 33
    2.4.1 Basic principles 34
    2.4.2 Instrumentation and methods 39
    2.4.3 Strengths and limitations 45
    2.4.4 General areas of application 46
2.5 Overall future directions and developments 48
References 49

3	**Solid phase extraction (SPE) in organic analysis** A. J. HANDLEY AND R. D. MCDOWALL	**54**

3.1 Introduction 54
3.2 The background and principles of solid phase extraction 54
3.3 Instrumentation for solid phase extraction 58
3.4 The solid phase extraction column/cartridge and disk 60
    3.4.1 The SPE column/cartridge 60
    3.4.2 The SPE disk 61
3.5 Solid phase extraction sorbents 62
3.6 Strategies for solid phase extraction 64
    3.6.1 The SPE process 64
    3.6.2 Method development in SPE 66
3.7 Applications of solid phase extraction 68
3.8 New technologies in solid phase extraction 70
    3.8.1 SPE phases 70
    3.8.2 SPE formats 71
    3.8.3 Linkage of SPE to other analytical techniques 71

| | References | 72 |

4 Solid phase microextraction (SPME) and membrane extraction with a sorbent interface (MESI) in organic analysis   75
Y. LUO AND J. PAWLISZYN

4.1	Introduction	75
4.2	Solid Phase Microextraction (SPME)	76
	4.2.1 Introduction	76
	4.2.2 Theoretical aspects of SPME optimization and calibration	81
4.3	Membrane Extraction with Sorbent Interface (MESI)	86
	4.3.1 Introduction	86
	4.3.2 Theoretical aspects of the membrane extraction process	90
	4.3.3 Practical aspects of MESI analysis	93
Acknowledgements		98
References		98

5 Supercritical fluid extraction in organic analysis   100
H.-G. JANSSEN AND X. LOU

5.1	Introduction	100
5.2	Properties of supercritical fluids	102
5.3	Modifiers or co-solvents	107
5.4	Instrumentation	111
	5.4.1 Solvent delivery pumps and ovens	111
	5.4.2 Extraction cells and cell-packing procedures in SFE	113
	5.4.3 Flow restrictors	115
	5.4.4 Solute collection	116
5.5	Method development	118
	5.5.1 Kinetic models for SFE extraction	119
	5.5.2 Sample morphology	121
	5.5.3 Sample size	121
	5.5.4 Selection of the supercritical fluid	122
	5.5.5 Pressure	123
	5.5.6 Temperature	124
	5.5.7 Flow rate and extraction time	125
	5.5.8 Collection	125
	5.5.9 Optimizing and validating SFE methods	126
	5.5.10 Using SFC retention data for SFE method development	129
5.6	General areas of application	129
	5.6.1 SFE in environmental extractions	130
	5.6.2 Applications of SFE in food analysis	133
	5.6.3 Extractions of polymeric materials	135
5.7	Future directions and developments	138
References		140

6 Pressurized fluid extraction (PFE) in organic analysis   146
J. L. EZZELL

6.1	Introduction	146

	6.2	Basic principles		147
	6.3	Instrumentation		148
	6.4	PFE methods development		150
		6.4.1	Sample preparation	150
		6.4.2	Grinding	150
		6.4.3	Dispersing	151
		6.4.4	Drying	151
		6.4.5	Extraction parameters	151
		6.4.6	Method validation	154
		6.4.7	Selectivity	155
	6.5	Strengths/limitations		155
	6.6	Areas of application		157
		6.6.1	Environmental	157
		6.6.2	Polymers	159
		6.6.3	Foods	162
		6.6.4	Pharmaceuticals and natural products	163
	References			164

7 Microwave-assisted solvent extraction in organic analysis — 166
J. R. DEAN, L. FITZPATRICK AND C. HESLOP

7.1	Introduction		166
7.2	Microwave interaction with matter		166
	7.2.1	Choice of reagents	168
	7.2.2	Solvent effects	168
7.3	MAE systems		169
7.4	Heating methods		171
7.5	Application of MAE for environmental analysis		172
	7.5.1	Polycyclic aromatic hydrocarbons (PAHs)	174
	7.5.2	Pesticides	175
	7.5.3	Polychlorinated biphenyls	187
	7.5.4	Phenols	187
	7.5.5	Phthalate esters	189
7.6	Microwave-assisted solid phase extraction		189
7.7	Gas-phase microwave-assisted extraction		190
7.8	Future prospects for MAE		190
References			191
Appendix			193

8 Biological/pharmaceutical applications — 194
D. STEVENSON, S. MILLER AND I. D. WILSON

8.1	Introduction		194
8.2	Techniques in common use		197
	8.2.1	Protein preparation	198
	8.2.2	Liquid-liquid extraction	199
	8.2.3	Solid phase extraction	202
	8.2.4	Homogenisation/hydrolysis of solids	207
	8.2.5	Hydrolysis of conjugates	207
	8.2.6	Dialysis	208
	8.2.7	Column-switching	208

	8.2.8 Restricted Access Media (RAM) columns	209
	8.2.9 Volume reduction	209
	8.2.10 Derivatisation	210
	8.2.11 Headspace analysis	212
	8.2.12 Automation of sample preparation	213
8.3	Applications	215
8.4	Factors influencing choice of method	216
8.5	Future prospects	216
Bibliography		217
References		218

9 Polymers and polymer additives   221
H. J. VANDENBURG AND A. A. CLIFFORD

9.1	Introduction	221
9.2	Dissolution/reprecipitation	221
9.3	Liquid/solid extraction	222
9.4	Traditional methods of solid/liquid extraction	224
	9.4.1 Soxhlet extraction	224
	9.4.2 Boiling under reflux	224
	9.4.3 Sonication	225
9.5	High pressure solid/liquid extraction methods	226
	9.5.1 Supercritical fluid extraction (SFE)	226
	9.5.2 Accelerated solvent extraction (ASE)	233
	9.5.3 Microwave-assisted extraction	237
9.6	Comparison of techniques	239
References		240

10 Environmental applications   243
P. SANDRA, F. DAVID, E. BALTUSSEN AND T. DE SMAELE

10.1	Introduction	243
10.2	Key extraction methods	244
	10.2.1 Introduction	244
	10.2.2 Air and gaseous samples	244
	10.2.3 Water and liquid samples	247
	10.2.4 Soil, sludge, sediment and solid samples	250
10.3	Instrumentation	253
	10.3.1 Introduction	253
	10.3.2 Gas chromatography (GC)	254
	10.3.3 Liquid chromatography (LC)	255
10.4	Selected applications	256
	10.4.1 Dynamic sorptive extraction of pollutants in air	256
	10.4.2 Dynamic sorptive extraction of nicotine in hospital air	259
	10.4.3 Micro-liquid-liquid extraction of trihalomethanes in drinking water	261
	10.4.4 Static sorptive extraction of pesticides from water samples	266
	10.4.5 Static and dynamic sorptive extraction of phenols in water samples	268
	10.4.6 Solid phase extraction of triazines from water samples on Empore disk cartridges followed by large volume injection	278

10.4.7	Solid phase extraction of pesticides from water samples followed by supercritical fluid desorption and supercritical fluid chromatography	281
10.4.8	Gas phase extraction of PAHs and PCBs from soil samples	286
10.4.9	Static sorptive extraction of organometallics from environmental samples	288
10.4.10	Supercritical fluid extraction of PCBs from oils and fats	297
10.4.11	Supercritical fluid extraction of pesticides in orange juices	300
References		302

Index   **305**

1 Introduction
Alan J. Handley

Organic analysis plays an important role in the development of current science as it underpins many of the key areas of technology, such as pharmaceutical and biological science, polymer technology, food technology and agrochemicals, and it can provide the means of monitoring and regulating such industries. The dictionary describes organic analysis as the resolution of a chemical compound into its proximate or ultimate parts, and the determination of its elements or the foreign components it may contain. This process can be designed to provide qualitative data, but it can also lead to the need for quantitative measurement.

Analysis is comprised of three key processes: sample preparation, sample analysis and data handling and interpretation—often described as the 'Analysis Cycle' (Figure 1.1).

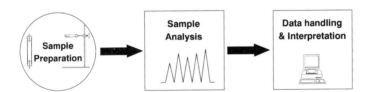

Figure 1.1 The Analysis Cycle.

Although some samples are inherently ready for analysis, most require some form of pretreatment before they can be presented to the analytical technique of choice. The pretreatment will depend on the physical requirements of the instrument (liquid, solid or gaseous sample), the scope of the measurement (ppb, ppm, percentage levels of the determinants), and the complexity of the sample. Sample pretreatment in its simplest form can involve dilution, filtration, evaporation or centrifugation. However, with the continuing trend to examine more complex samples and samples with constituent components at lower and lower levels, sample extraction is becoming increasingly important as a preparation step. This is evident if we examine the main areas from which we obtain or generate samples for organic analysis.

Product development—the pharmaceutical industry has seen an explosion of newer products requiring development and testing, with analytical assays having to be closely regulated and determinations made in very complex matrices.

Product formulation—the pharmaceutical, food and agrochemical industries are required to control and monitor lower and lower levels of by-products in their formulations which, in themselves, are becoming more complex in nature.

Product production—increasing pressures to control plant effluents, together with the ability of newer analytical technologies to measure lower and lower compound levels, has resulted in increased regulatory restrictions on compounds previously considered to be nondetectable.

All these factors result in the need for more sample pretreatment/extraction prior to analysis, to enable accredited analytical data to be obtained. This may involve either sample clean-up, sample concentration, or changing the physical form of the sample to make it more amenable to the analytical technique of choice (Figure 1.2).

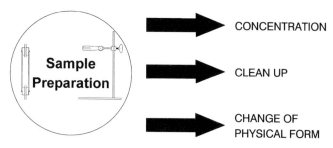

Figure 1.2 The role of extraction in sample preparation.

Sample extraction encompasses all those steps necessary to prepare the original sample for determination by the chosen method. Strategically, the extraction method chosen can influence the entire process, and will depend on: the chemistry of the analyte(s); the nature of the matrix; the objectives of the analysis; and the limit of determination. The relative importance of these factors will vary depending on the measurement requirement.

Many extraction practices, however, are based on classical methodologies of liquid/liquid or liquid/solid extraction, using practices (separating funnels and Soxhlet extractors) which have not changed appreciably for the past one hundred years. The most commonly used technique has been liquid extraction, based on the observation that organic compounds can usually be removed from a matrix by extraction with a water-immiscible solvent. This method relies largely on the relative solubility, diffusion, partitioning, viscosity or surface tension of the component(s) of interest and the solvent. At first sight, this may appear simple and straightforward but, to achieve high recoveries, it is often necessary to repeat the extraction process several times and then to bulk the fractions together prior to analysis. This is also true of solid/liquid extraction, where samples have to be continually extracted—in some cases for days—to obtain the desired efficiency of extraction. So, while classical extraction procedures can be very effective, they can also be extremely labour-intensive and time-consuming, and can require large amounts of solvent.

Two principal factors have been responsible for speeding up developments in the area of extraction:

(i) The need for faster, high throughput analysis. If we look back at the Analysis Cycle, there have been tremendous advances in the automation capabilities of the equipment for both analytical measurement and data-handling, to the extent that, for a complete qualitative and quantitative analysis, the analysis times have decreased from hours or sometimes days. Thus, in many cases, sample preparation or extraction is now the major rate-determining step in the analysis.

(ii) The concerns about the hazardous nature of many of the commonly used solvents, the problems and cost of disposal, and the associated question of solvent emissions during the concentration of large volumes of sample extracts.

Recent years have seen considerable efforts to develop alternative and improved methods of extraction, and to make them commercially available. This has resulted in a number of systems being developed commercially for both liquid/liquid and solid/liquid extraction (Table 1.1).

Table 1.1 Extraction technologies

Classical	Primary sample matrix	Primary extraction phase	New technologies
Solvent extraction	Liquid	Liquid	Solid phase extraction
Membrane extraction			Solid phase micro-extraction
Soxhlet	Solid	Liquid	Accelerated or pressurised fluid extraction, Microwave-assisted solvent extraction, Supercritical fluid extraction

Some, such as Supercritical Fluid, Accelerated or Pressurised, and Microwave-assisted Extraction, use different driving conditions to speed up the process, while others use alternative technologies, such as Solid Phase and Solid Phase Micro-extraction, to reduce solvent volumes. Other approaches have involved the evolution of classical methodology, with the launch of the automated Soxhlet and countercurrent extraction systems.

Such developments have resulted in systems now capable of providing: faster extractions and systems which are easier to automate; completeness of extraction, with recoveries equal to or exceeding those of older methods; analyte selectivity; robust and reproducible methodology, equal to or exceeding that of older methods; elimination of the use of some of the more hazardous solvents, by the selection of different modes of extraction (pressure, temperature and density); reduction of solvent volumes and, hence, reduction in the cost of operation (solvent use and disposal); and ease of coupling to other analytical instrumentation.

The chapters that follow describe the various technologies in full, together with the processes, the equipment, the protocols for method development, and the many applications.

2 Solvent and membrane extraction in organic analysis

Alexis J. Holden

2.1 Introduction

There are three main reasons for using solvent and/or membrane extraction: (i) to isolate a component or analyte of interest; (ii) to remove potential interferents from a matrix; and (iii) to preconcentrate an analyte prior to measurement. This chapter will demonstrate the variety of methods available to perform these three objectives in matrices such as air, water, soil and foodstuffs. Table 2.1 provides a summary of some of the applications of the various techniques of solvent and membrane extraction.

Apart from the basic extraction of material, there are alternative sampling techniques which encompass the theory of solvent extraction and, in some cases, filtration. Such techniques are denuders and impregnated filters, both of which will also be discussed in this chapter.

2.2 Solvent extraction

Solvent extraction is defined as the process of separating one constituent from a mixture by dissolving it into a solvent in which it is soluble but in which the other constituents of the mixture are not (Isaacs *et al.*, 1989), or are at least less soluble. The separation process may involve a mixture composed of: (i) two or more solids; (ii) a solid and a liquid; or (iii) two or more liquids (Lewis, 1993). If the mixture is composed of two or more solids, then the extraction process is performed by the process of percolation, where a solvent is passed through the mixed solids and, thus, the analyte which is soluble in the solvent is extracted into that solvent. If the solvent used is water, then this process can be termed leaching (Thorpe and Whitely, 1940). When the term solvent extraction is used, it is usually taken as meaning extraction involving two or more liquids, i.e. liquid-liquid extraction. In liquid-liquid extraction, the solution containing the desired constituent must be immiscible with the liquid used to extract the desired constituent (Isaacs *et al.*, 1989). When the extraction process has occurred, the phase which contains the extracted analyte is known as the extract phase, while the sample from which the analyte has been removed is named the refined phase (Ulicky and Kemp, 1992).

Table 2.1 Examples of the application of solvent and membrane extraction techniques in organic analysis

Extraction method	Analyte	Matrix	Comments	Reference
IMF	Acetaldehyde	Indoor air	Impregnated with 2,4-dinitrophenylhydrazine	Lindahl et al., 1996
F, S	45 PAHs	Workplace air	Store filter and adsorbent together before extraction	Notø et al., 1996
D, F	Nicotine	Air (gases and particulates)	Samplers coated with benzenesulphonic acid, extracted with NaOH	Häger and Niessner, 1996
I, SS	Aldehydes and ketones	Air (gas stoves, car exhausts, cigarette smoke, restaurants)		Grömping and Cammann, 1996
SS, B, I	Cyclic OAAs	Air	A review	Jönsson et al., 1996
I, IMF	Isocyanates	Paint-spraying environment	Comparison of solid sample and impinger systems containing 1-(2-methoxyphenyl)piperazine	Maitre et al., 1996
I	Isocyanates	Air	Extracting solvent, dibutylamine	Spanne et al., 1996
I, F	PAHs	Workplace atmospheres	Extracting solution—methoxyethanol	Hiel et al., 1995
I	Acidic gases	Flue gas	Extrating solution—PR-DEAE sephadex, cellulose powder and water	Takeshita et al., 1995
CD	Hydrocarbons	Air		Risse et al., 1996

D	Carbonyl compounds	Stack gas	Coated with acidified 2,4-dinitrophenylhydrazine	Kallinger and Niessner, 1997
D	Gaseous species	Various	Review	Ali et al., 1989
D	Aromatic hydrocarbons	Workplace air	Coated with charcoal paper, compares different denuder designs	Risse et al., 1996
AD	Nicotine, 3-ethenylpyridine	Indoor environments		Bertoni et al., 1996
AD	PAHs	Indoor air, environmental tobacco smoke	Coated with 1% benzenesulphonic acid in methanol	Gundel et al., 1995
AD	Nitrogen dioxide, peroxyacetyl nitrate	Ambient atmosphere	Adsorbent, XAD-4	De Santis et al., 1996
AD	2,4-dinitrophenol vapour	Air	Active carbon, preparation explained	Oms-Molla and Klockow, 1995
D	2,4-dinitrophenol vapour	Air	Coated with NaOH-Ca(OH)$_2$	Oms-Molla and Klockow, 1995
I	PAHs	Air	Extraction solution 0.1 M NaOH	Coutant et al., 1992
AD	PAHs	Soil/compost		Wischmann et al., 1996
BE	Petrolatum products	OSEs	OSEs produce hazing and gelling	Spangler and Sidhom, 1996
S	PACs	Foodstuffs		Guillén, 1994
S	Organic chemicals	Soil	Solvent used to mimic bioavailability	Kelsey et al., 1997

Table 2.1 (Continued)

Extraction method	Analyte	Matrix	Comments	Reference
BE	1-naphthol	Soil		Hancock and Dean, 1997
SE, UF	High molecular mass fractions of lignin and humic compounds	Sediments and surface waters		Hyötyläinen et al., 1998
	PCDDs, PCDFs	Muscle and liver of black cormorants	—	Falandysz et al., 1996
CFF, MF, UF	TOC	Municipal textile effluents	Better removal of TOC than orthodox methods	Malpei et al., 1997
UF	Dissolved organic carbon, organic pollutants, triazines	Natural waters		Ludwig et al., 1997
		Oils		Martinez et al., 1995
HF	VOCs	Air emissions	On-line preconcentration and analysis using one or more membranes	Mitra et al., 1996
NF	NOM	Drinking water		Braghetta et al., 1997
PV	1-butanol	Fermentation broths	Various membrane materials tested	Favre et al., 1996
PV	Dairy aroma compounds	Aqueous solutions		Baudot and Marin, 1996
SM	Phenols	Crude oils	Membrane coupled directly to chromatographic instrumentation	Garcia Sanchez et al., 1997
SLM	Amino acids		Membrane containing diethylhexyl phosphoric acid	Wieczorek et al., 1997

UF	Oxygenates	Comparison of flat-sheet SLM and hollow-fibre membranes	Ozadali et al., 1996
HM	Citrus oils		Brose et al., 1995
CFF	Organic colloids	Seawater	Gustafsson et al., 1996
NF	Organic compounds	Bleaching pulp, paper effluents Compares different designs	Afonso and de Pinho, 1997

Abbreviations: S, solvent; D, denuder; F, filter; I, impinger; SS, solid sorbent; B, bubbler; CD, charcoal denuder; AD, annular denuder; BE, batch extraction; SE, solvent extraction; UF, ultrafiltration; CFF, crossflow filtration; MF, microfiltration; HF, hollow fibre; NF, nanofiltration; PV, pervaporation; SM, silicone membrane; SLM, supported liquid membrane; HM, hydrophilic membrane; PAH, polycyclic aromatic hydrocarbon; PAC, polyaromatic carbon; NOM, natural organic matter; OSE, organic solvent extractable; OAA, organic acid anhydride; PCDD, polychlorinated dibenzo-p-dioxin; PCDF, polychlorinated dibenzofurans; VOC, volatile organic compound; TOC, total organic carbon; IMF, impregnated membrane filtration; PR-DEAE, PR-diethylaminoethyl; XAD-4, a polymeric, hydrophobic adsorbent.

2.2.1 Basic principles

The theory of equilibrium and the extraction of solutes into various phases, whether liquid or vapour, are thoroughly described in terms of physical chemistry in a number of publications (Schwarzenbach *et al.*, 1993; Atkins, 1986; Warn, 1969) and, as such, they will be described only briefly in this chapter.

The extraction of an analyte from one phase into a second phase is dependent upon two main factors: solubility and equilibrium. The principle by which solvent extraction is successful is that 'like dissolves like'. That is, to remove a polar solute from a solution a polar solvent should be used and to remove a nonpolar solvent from a solution a nonpolar solvent should be used. Qualitative predictions can be made on the likely success of an extraction by considering the polarity of the analyte of interest and of the two solvents used. Uncharged solutes are more easily extracted into nonpolar organic solvents, and the less polar the solute the more efficient the extraction process. If the solute is charged, then it is usually best to form an ion-pair with a counter ion and extract the newly formed neutral complex into a nonpolar solvent. The major problem in extracting polar solutes into polar solvents is the miscibility of polar solvents with water, which is the main matrix for many samples. The chemical form of an analyte has a fundamental effect on the efficiency of an extraction and is dependent upon the equilibria of the system within which it is contained. The successful selectivity of a solvent extraction process relies upon the judicious manipulation of the chemical equilibria to produce a species of solute which is preferentially soluble in one of the two solvent phases, usually either organic or aqueous.

To identify which solvent performs best in which system, a number of chemical properties can be calculated to determine the efficiency and success of an extraction:

(i) The distribution/partition coefficient (K_D)
Generally the extraction process involves the separation of a dissolved substance, known as the solute, from a liquid, solid or gaseous sample by using a suitable solvent. This process is reliant upon the relationship described by Nernst's distribution law, which is also known as the partition law.

The partition law, which describes the distribution coefficient (K_D), is a relationship in which a dissolved substance (solute) is distributed between two immiscible phases. The two immiscible phases can be a solid and a liquid, a liquid and a liquid, or a gas and a liquid. The two phases interact either by shaking them together or spraying/bubbling one into the other, and the solute (S) will distribute itself between the two phases depending on the affinity/solubility of the solute for each phase. The ratio of the solute concentration in each phase will be constant:

$$K_D = [S]_E/[S]_O \qquad (2.1)$$

where, K_D is the distribution coefficient, $[S]_E$ represents the concentration of the solute in one of the phases, and $[S]_O$ represents the concentration of the solute in the second phase. Usually, $[S]_O$ would represent the sample which contained the solute originally and $[S]_E$ would represent the solvent which has been used to extract the solute from the original sample matrix. Therefore, the greater the solubility of the solute in the extracting solvent the greater the value of K_D. A K_D value is characteristic for a given solute and solvent system and is dependent on temperature. When the sample contains several components, each dissolved substance distributes itself between the solvents according to their own distribution coefficient.

(ii) The distribution ratio (D)

The distribution ratio (D) is the ratio of the concentration of all the species of the solute in each phase. For example, with a substance, HS, which may have partially dissociated, the ratio would be:

$$D = [HS]_E/([HS]_O + [S^-]_O) \qquad (2.2)$$

The relationship between K_D and D is as follows. Assuming that one of the phases is an aqueous phase containing the analyte which is an acid, then the acidity constant (K_a) for the ionization of the acid in the aqueous phase is given by:

$$K_a = [H^+]_O[S^-]_O/[HS]_O \qquad (2.3)$$

Note, the subscript O, $[H^+]_O$, represents the species in the original sample and the subscript E, $[HS]_E$, represents the species in the extracting solvent. From Equation 2.3, it follows that:

$$[S^-]_O = K_a[HS]_O/[H^+]_O \qquad (2.4)$$

From Equation 2.1:

$$[HS]_E = K_D[HS]_O \qquad (2.5)$$

Substitution of Equations 2.4 and 2.5 into Equation 2.2 gives:

$$D = K_D[HS]_O/([HS]_O + (K_a[HS]_O/[H^+]_O)) \qquad (2.6)$$

which rearranges to produce:

$$D = K_D/(1 + (K_a/[H^+]_O)) \qquad (2.7)$$

The above equation suggests an approximation that when $[H^+]_O >>> K_a$, then $D = K_D$, and D is a maximum and if K_D was large then S (the required analyte) was extracted into the extracting solvent; and when $[H^+]_O <<< K_a$, then $D = K_D[H^+]_a/K_a$, which will be small and no extraction will occur.

By the nature of the extracting solvent and the solute extracted the solute may undergo polymerisation in the organic phase. An example of this situation is the extraction of a carboxylic acid into an organic solvent, such as benzene (Peters *et al.*, 1974). The extraction process is explained by a modification of Equation 2.1. The overall equilibrium is shown below. The acid will not dissociate in the aqueous phase if the pH is less than 2.

$$\text{HA} \rightleftharpoons \bigg| \text{HA} \rightleftharpoons (\text{HA})_2$$

Aqueous phase | Organic phase

The overall equilibrium is governed by the following two equations:

$$[K_D] = [HA]_E/[HA]_O \qquad (2.8)$$

$$[K_2] = [(HA)_E]_E/[HA]_E^2 \qquad (2.9)$$

and the distribution ratio is determined by:

$$[D] = ([HA]_E + 2[(HA)_2]_E/[HA]_O \qquad (2.10)$$

Substituting Equations 2.8 and 2.9 into Equation 2.10 gives:

$$D = K_D + 2K_2K_D^2[HA]_O \qquad (2.11)$$

So, from Equation 2.11, the distribution ratio is dependent upon the concentration of the total acid concentration in the original sample.

(iii) The partition isotherm

The shape of a partition isotherm can be used to deduce the behaviour of the solute in the extracting solvent. Figure 2.1 shows three different partition isotherms for solute behaviour. Line (a) represents the expected ideal behaviour when the distribution ratio remains constant. Line (b) represents the variation in the distribution ratio when the solute has polymerised, e.g. formed dimers, trimers or polymers, etc. Line (c) represents the distribution ratio when the solute is adsorbed onto a surface of finite area, e.g. membrane extraction when the concentration of the solute on the surface approaches a monolayer coverage.

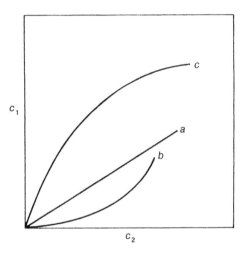

Figure 2.1 Three examples of partition isotherms for solute behaviour. Line (a) represents the expected behaviour when the distribution ratio remains constant. Line (b) represents the variation in the distribution ratio when the solute has polymerized. Line (c) represents the distribution ratio when the solute is adsorbed onto a surface of a finite area. (Reproduced with permission from Peters *et al.* (1974) *Chemical Separations and Measurements: Theory and Practice of Analytical Chemistry*, W.B. Saunders Co., London, UK, p. 489).

(iv) The percentage extracted

Whilst the distribution ratio (D) is independent of the ratio of the volumes of the two phases used, the concentration of the solute extracted is not. The fraction of the solute extracted is equal to the concentration of the solute in the extracting phase divided by the total concentration of the solute. Therefore, the percentage extracted can be calculated using Equation 2.12:

$$\%E = ([S]_E V_E / ([S]_E V_E + [S]_O V_O)) \times 100 \qquad (2.12)$$

where, V_E and V_O are the volumes of the extracting phase and the sample phase, respectively. By considering Equation 2.12, the relationship between the percentage extracted and the distribution ratio is shown by:

$$\%E = 100D/(D + (V_O/V_E)) \qquad (2.13)$$

If equal volumes were used, so $V_O = V_E$, then:

$$\%E = 100D/(D + 1) \qquad (2.14)$$

As stated previously, one of the reasons for using solvent extraction is to remove an analyte from its matrix. To do this, there may be a temptation to keep extracting a sample until there is virtually no analyte remaining in the matrix. Equation 2.15 shows that there is little benefit in performing more than five sequential extractions on the same sample.

$$[A_{aq}]_n = ((V_{aq}/(V_{org}K + V_{aq}))_n [A_{aq}]_0 \qquad (2.15)$$

where, $[A_{aq}]_n$ is the concentration of A remaining in the aqueous solution after extracting V_{aq} cm^3 of the solution having an original concentration of $[A_{aq}]_0$ with n portions of the organic solvent, each having a volume of V_{org}. Therefore, Equation 2.15 proves that it is always better to use several small portions of solvent (e.g. 5 × 20 cm^3) to extract a sample than to extract with one large portion (e.g. 1 × 100 cm^3).

2.2.2 Instrumentation and procedures

When using any solvent extraction system, one of the most important decisions is the selection of the solvent to be used. The properties which should be considered when choosing the appropriate solvent are: selectivity; distribution coefficients; insolubility; recoverability; density; interfacial tension; chemical reactivity; viscosity; vapour pressure; freezing point; safety; and cost (Price, 1997). Table 2.2 presents an example of the typical processes which are involved in a solvent extraction process. Table 2.3 provides examples of the solvents which may be used in such procedures. Figure 2.2 provides a schematic representation of the four levels at which solvent extraction can be performed: (i) single step extraction; (ii) multistep extraction; (iii) countercurrent extraction; and (iv) cross-current extraction. Table 2.4 summarizes when each extraction method should be used.

Table 2.2 Typical processes in a solvent extraction procedure using, as an example, an air filter which has collected polycyclic aromatic hydrocarbons

Steps		Example*
1.	Prepare a solution of the solutes to be separated. If the solute is in a solid matrix, then a solution may be prepared.	Put air filter into a solution of cyclohexane and place in an ultrasonic bath. The analyte, **PAHs**, will extract into the cyclohexane solution along with any polar compounds.
2.	Adjust solution to ensure most efficient extraction conditions. This may involve preparing derivatives of the solute or adjusting the chemistry of the solution (e.g. pH) to maximise the difference in solubility of the solute between the two solvents.	
3.	Add a second solvent to the system which is immiscible with the initial solvent that the solute is dissolved in.	Add DMF containing 3% water, which will extract the polar compounds but not the PAHs.
4.	Shake the mixture in a sealed, stoppered container.	Shake
5.	Allow the mixture to stand and the two phases to separate into two distinctive layers.	
6.	Collect each layer separately and analyse as required.	Collect the cyclohexane layer and dry with anhydrous sodium sulphate. This fraction will contain the PAHs.

*Modified from Noto *et al.*, 1996.
Abbreviations: PAH, polycyclic aromatic hydrocarbon; DMF: dimethyl formaldehyde.

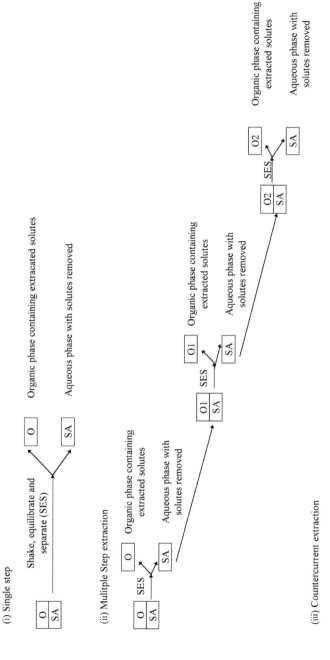

Figure 2.2 Schematic representation of the four levels at which solvent extraction can be performed.

SOLVENT AND MEMBRANE EXTRACTION 17

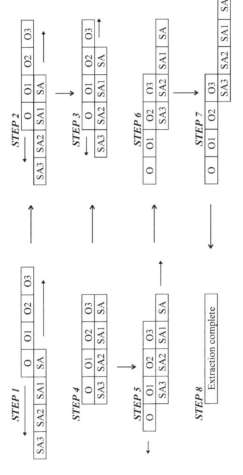

Figure 2.2 (Continued).

Table 2.3 Examples of the solvents used in solvent extraction procedures

Aqueous phase solvents	Organic phase solvents
(i) Distilled deionised water	Toluene and isomeric xylenes
(ii) Aqueous buffer solution to a specific pH	Di-, tri-, tetrachloromethane
(iii) Aqueous solution of electrolytes	Ethers
(iv) Aqueous solution containing a complexing agent	Esters
(v) Aqueous acid	Specific ketones (those immiscible with water)
(vi) Aqueous alkali	Alcohols of high molecular mass
(vii) Combination of two or more from (i) to (vi)	Aliphatic hydrocarbons

2.2.2.1 Simple extractions, the separating funnel and the bubbler/impinger system

The most simple apparatus required for liquid-liquid extraction is the separating funnel (Figure 2.3A). The sample from which the solute is to be removed and the liquid into which the solute is to be extracted are both placed in the separating funnel. The liquids are then shaken together, carefully releasing the pressure when necessary by inverting the funnel and opening the tap. The liquid phases are then left to settle, with the denser phase on the bottom. Before opening the tap to collect the separated phases, the top of the separating funnel should be removed! By opening the tap of the funnel, the phases can then be collected separately. If the extraction procedure is performed only once, then it is termed a single extraction method. The extraction process may be repeated up to five times using fresh solvent each time to extract the majority of the solute from the solution; this is termed multiple extraction. The separate extracts would then be combined and, if the concentration of the analyte in the final solution was too low, some of the solvent could be evaporated to reduce the volume in which the analyte is dissolved, thus increasing the analyte concentration. It may be necessary to remove and evaporate all of the solvent due to incompatibility with the next stage in the analysis and, hence, the analyte would be 'taken-up' in a second matrix, e.g. acid. The procedure of solvent extraction is simple, rapid and quantitative, requiring the minimum of apparatus. This type of extraction is employed when the distribution ratio is favourable for one solute species in a solution ($D > 5$) and unfavourable for the other components of that solution ($D < 0.001$) (Skoog et al., 1992).

Whatman® have developed the VectaSep CLE®, a liquid-liquid extraction system, which uses a two-step centrifugation method instead of the standard separating funnel and allows the extraction of a number of samples simultaneously.

SOLVENT AND MEMBRANE EXTRACTION

Table 2.4 Summary of the different extraction procedures to use

Properties of sample and analyte	Appropriate solvent extraction system
Analyte, D > 5 and matrix D < 0.001	Simple separating funnel extraction using either a single extraction or multiple extractions
Analyte, D < 0.001 and matrix D < 0	Exhaustive extractions, Soxhlet extraction
D of analyte similar to D of matrix	Countercurrent extraction
Analyte more likely to be in the vapour phase	Bubbler extraction system
Analyte more likely to be in a solid, particulate form	Impinger extraction system

Abbreviation: D, distribution ratio.

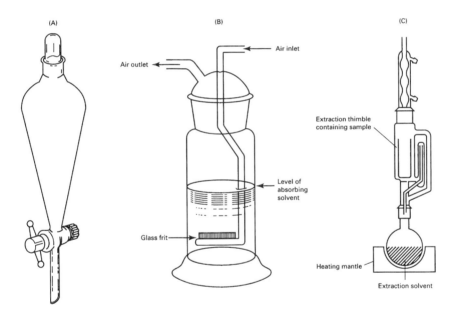

Figure 2.3 A, Separating funnel. B, An impinger/bubbler system. (Reproduced with permission from Manahan (1991) *Environmental Chemistry*, 5th edn, Lewis Publishers, Michigan, USA, p. 460). C, A Soxhlet system. (Reproduced from Reeve (1994) *Environmental Analysis*, John Wiley and Sons, Chichester, UK, p. 150).

A typical method by which gases can be solvent extracted is the use of a bubbler or impinger system. Some workers have stated that a bubbler is synonymous with an 'impinger', and in this chapter they will be treated as essentially the same technique. However, it must be noted that the impinger technique can also refer to the collection and separation of particulate matter. Streicher and co-workers described the difference between a bubbler and an impinger by the way the air is introduced into the liquid (Streicher et al., 1994). In a bubbler system, the airstream forms very small bubbles after passing through sintered glass. Due to the small bubbles formed and hence greater surface:volume ratio, there is a collection efficiency which is better than that experienced by the impinger system, which does not use sintered glass. Jönsson and co-workers decided whether to employ an impinger or a bubbler depending upon the phase composition of the air matrix to be sampled. If more particulate material was expected then an impinger was used but if the analyte was more likely to exist in the vapour phase then a bubbler was used (Jönsson et al., 1996). The two types of impingers as described and used by the Occupational Safety and Health Act (OSHA) and the National Institute for Occupational Safety and Health (NIOSH) are the midget impinger, which is a smaller version of the Greenburg-Smith version, and the fritted glass bubbler.

The basic operation of a bubbler/impinger system involves the pumping of a gaseous sample through a liquid phase; the analyte of interest, which should have a greater affinity for the liquid phase, will dissolve into the liquid phase which will subsequently be analysed for the analyte of interest. A bubbler/impinger system is presented in Figure 2.3B. Lin and co-workers (1997) described the mechanisms which affect the efficiency of an impinger system (Figure 2.4). The problems associated with impaction, bounce and reaerosolization and their effect on the efficiency of impinger systems have been discussed previously (Grinshpun et al., 1997). Lin and co-workers also investigated the effect of sampling time on the collection efficiency of all-glass impingers (Lin et al., 1997). They reported that the critical dimension for the performance of an impinger was the inner diameter of the impingement nozzle, which helped to explain why various impingers performed differently; the inner diameter of the impingement nozzle was variable. They reported that about 1% of the solution volume is evaporated per minute, therefore as sampling progresses the extraction efficiency is decreased.

2.2.2.2 Exhaustive extractions and Soxhlet apparatus
When the distribution ratio of the solute of interest is considered unfavourable ($D < 0.001$) and it is to be extracted from other components of a sample solution with an even less favourable distribution ratio ($D \rightarrow 0$), then an exhaustive extraction procedure should be used. The main apparatus used is a Soxhlet extractor (Figure 2.3C), which enables the organic solvent to be automatically distilled, condensed and passed through the aqueous solvent or solid many hundreds of times. This process is equivalent to hundreds of simple extractions but will be completed in much less time, approximately one hour (Skoog et al., 1992). This process may also be termed reflux

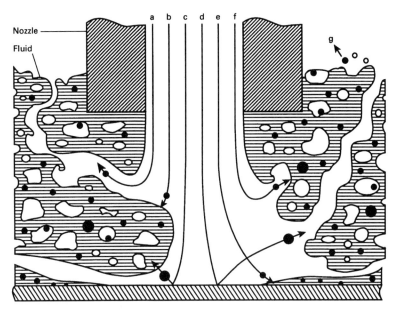

Figure 2.4 Illustration of the mechanisms that affect the collection efficiency of impingers. Particles are collected into the liquid (b); indirect impingement into the liquid after bounce from the bottom surface (c); impingement into the liquid layer formed on the bottom surface (e); and diffusion into the liquid (f). Particles are not collected due to insufficient inertia (a); or escape with the effluent airflow after bounce from the bottom surface (d). Collected particles may be reaerosolized when contained in a bursting bubbles (g). (Reproduced with permission from Lin et al. (1997) Effect of sampling time on the collection efficiency of all-glass impingers. *AIHA*. **58**, p. 486).

extraction. Soxhlet extraction is ideal for solid samples and the solutes required are extracted via the process of percolation. Soxhlet extraction is described in more detail in Chapter 6.

2.2.2.3 Countercurrent extraction and the Craig extraction apparatus
When the distribution ratios of components are very similar, then countercurrent extraction procedures will be successful. This procedure differs from exhaustive extractions in that it uses fresh portions both of the organic extracting phase and the aqueous phase sample containing the solute to be extracted. Figure 2.5 presents the typical apparatus for the countercurrent extraction process and the procedural steps are described in Figure 2.5 and Table 2.5. The operation of the Craig extraction apparatus is described in Table 2.6, with reference to Table 2.5. This procedure is discussed in more detail in Chapter 6.

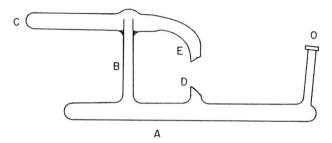

Figure 2.5 Apparatus for countercurrent extraction. The operation of the Craig extraction apparatus is described in Table 2.6, with reference to Table 2.5. A to O are used to explain the movement of the apparatus through its operation, as described in Table 2.6. (Reproduced from Peters et al. (1974) *Chemical Separations and Measurements: Theory and Practice of Analytical Chemistry*. W.B. Saunders Co., London, p. 506).

Table 2.5 Procedural steps involved in countercurrent extraction

Step	Procedure					
1	Fresh organic phase (O) added to aqueous phase containing solutes (SA)	O / SA				
2	Fresh organic phase (O1) added to original aqueous phase containing solutes (SA); and O added to fresh aqueous phase containing no solutes (A1)	O1 / SA	O / A1			
3	Fresh organic phase (O2) added to original aqueous phase containing solutes; and O1 added to fresh aqueous phase containing no solutes (A1); and O added to fresh aqueous phase containing no solutes (A2)	O2 / SA	O1 / A1	O / A2		
4	Fresh organic phase (O3) added to original aqueous phase containing solutes; and O2 added to fresh aqueous phase containing no solutes (A1); and O1 added to fresh aqueous phase containing no solutes (A2); and O added to fresh aqueous phase containing no solutes (A3)	O3 / SA	O2 / A1	O1 / A2	O / A3	
5	Fresh organic phase (On) added to original aqueous phase containing solutes; and O3 added to fresh aqueous phase containing no solutes (A1); and O2 added to fresh aqueous phase containing no solutes (A2); and O1 added to fresh aqueous phase containing no solutes (A3); and O added to aqueous phase containing no solutes (An)	On / SA	O3 / A1	O2 / A2	O1 / A3	O / An

2.2.2.4 Crosscurrent extraction

This is a development of countercurrent extraction, in which the two phases involved in the extraction process are moving in opposite directions to each other, and is schematically presented in Figure 2.2.

Table 2.6 Operation of the Craig extraction apparatus

Step	Procedure	Precaution
1	A known volume of the solvent (SA) containing the solutes to be extracted is added to compartment A of the first tube via opening O	When the Craig tube is rotated counterclockwise about 100 degrees, none of the solvent, SA, should run through side-arm B
2	A known volume of the extracting solvent (O) is also introduced to compartment A of the first tube	When the Craig tube is rotated counterclockwise about 100 degrees, none of the solvent, O, should run through side-arm D
3	A known volume of the solvent (SA), which contains **none** of the solutes which are to be extracted is added to each of the other Craig tubes in the extraction apparatus	When the Craig tubes are rotated counterclockwise about 100 degrees, none of the solvent, SA, should run through the respective side-arm B
4	The Craig apparatus is gently rocked back and forth; similar to the use of a separating funnel, to allow equilibrium to be attained	The Craig apparatus should be kept horizontal
5	The Craig apparatus is rotated counterclockwise through 100 degrees; this causes the solvent (O) containing the solutes to flow through side-arm B into compartment C	
6	The Craig apparatus is returned to the horizontal position and the solvent (O), now containing the extracted solutes, will flow from compartment C through tube E into compartment A of the next Craig tube	
7	A fresh portion of the solvent (O) is added to compartment A of the first Craig tube	
8	Steps 4 to 6 are repeated	

2.2.2.5 Miscellaneous procedures

Backwashing—is the term given to the process of shaking the organic layer containing the extracted solute with the aqueous layer, which contains all reagents except the solute of interest. This process should remove any co-extracted species from the organic layer back into the aqueous layer.

Back extraction—occurs when the organic layer containing the extracted solute is shaken with the aqueous layer, which has been chemically modified, e.g. pH changed. The results will be the back extraction of co-extracted species from the organic layer back into the aqueous layer. This process may be referred to as a 'multistage extraction' and will be quite specific.

Evaporation—the organic layer containing the extracted solute can be warmed and the volume reduced due to the evaporation of the organic solvent. This would then cause an increase in the concentration of the extracted solute.

Drying—the organic layer containing the extracted solute may be filtered or placed over a drying agent, which would remove any traces of water from the organic phase.

2.2.3 Strengths and limitations

As the extraction efficiency is independent of the original concentration, solvent extraction is highly applicable for trace levels of analytes. Solvent extraction can be used either to preconcentrate a sample to ensure a more reliable and accurate result or to remove interferents, which may cause a problem with the final method of analyte determination. Moreover, due to the nature of many solvents, the extracted volume which contains the analyte of interest may be reduced by gentle heating to further increase the concentration effect. Simple solvent extraction is ideal for non-routine analysis but if multiple extractions are required then the Soxhlet system offers greater advantages, in that the process can be completed in a much quicker time and with less manual interaction. There may be problems with identifying the boundary of the two phases in the separating funnel due to the formation of an emulsion.

2.2.4 General applications

One of the main applications of solvent extraction is the removal of phenol from by-product water produced in coal-coking, petroleum-refining and chemical syntheses that involve phenol (Manahan, 1991). Solvent extraction has been employed to investigate particulate and gaseous polycyclicaromatic hydrocarbons (PAHs) in workplace air of electrode paste plants (Notø *et al.*, 1996). Two samplers were compared; the Institute of Occupational Medicine inhalable sampler and the Gelman total dust sampler. Both samplers had adsorbent behind the filter. The filter was used to isolate the dust particles, which were then extracted for PAHs using cyclohexane and dimethyl formaldehyde (DMF). The gaseous PAHs were extracted from the polymeric, hydrophic adsorbent, XAD-2, using dichloromethane. All samples were identified and quantified using gas chromatography with flame ionisation detection.

Hiel and co-workers also investigated PAHs in workplace atmospheres. They used solvent extraction as the end-step after employing a filter-impinger system (Hiel *et al.*, 1995). The workplace atmosphere was pumped through a glass wool filter and an impinger system containing methoxyethanol into which the PAHs would be extracted. The methoxyethanol solutions were then transferred to a separating funnel, and the PAHs were extracted three times into a cyclohexane methoxyethanol mix. The combined cyclohexane samples were then dried and analysed. Guillén (1994) reported the use of solvent extraction in the isolation of PAHs and polyaromatic carbons (PACs) in foodstuffs. PAHs and PACs are found both in processed and nonprocessed foods due to environmental contamination and procedures during the processing, preservation and packaging of foods which give rise to such compounds. When extracting low concentrations of PACs from foods, lipids, which are in high concentrations, are co-extracted; therefore, the process of extraction is multistepped using a variety of solvents.

Spangler and Sidhom (1996) used solvent extraction for the separation of active ingredients from samples of petroleum-containing pharmaceutical creams and ointments. They reported that extraction with methanol, acetonitrile and/or tetrahydrofuran will cause hydrocarbons to be co-extracted. The co-extracts can cause hazing and gelling and have the potential to block membrane filters and chromatographic columns. Some of the problems described were reduced if the extracts were chilled and filtered to remove the co-extracts.

Recently, a method was proposed for the use of bromine water for the extraction of organic mercury from hydrocarbons (Heyward *et al.*, 1997). The recoveries for dimethylmercury and diphenylmercury were: $98 \pm 5\%$ in heptane and $98 \pm 6\%$ in a condensate-heptane mixture for dimethylmercury; and $93 \pm 5\%$ in heptane and $95 \pm 5\%$ in a heptane-toluene-condensate mixture for diphenylmercury.

2.3 Other methods involving solvent extraction and/or filtration procedures

Sampling methods which require solvent extraction will be discussed in this section. Such methods include denuders and impregnated filters.

2.3.1 Basic principles

A schematic diagram of each of the techniques described below is presented in Figure 2.6, 2.7, 2.8 and 2.9.

Denuders
With the denuder technique, air containing the analyte to be sampled is sucked through a long cylindrical tube, which has been coated with a specific reagent. As the air passes through the tube, the analyte is

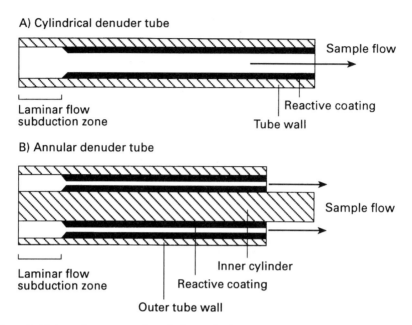

Figure 2.6 Schematic representation of: A, a cylinder denuder; and B, an annular denuder. (Taken from Ali *et al.* (1989) Denuder tubes for sampling gaseous species. *Analyst*, **114**, p. 760. Reproduced by permission of The Royal Society of Chemistry).

extracted from the sample by dissolving into the reagent coated onto the wall of the tube. For a quantitative analysis to be made, the lining of the denuder tube is extracted and analysed (although there are variations on this, as described in Section 2.3.2.1), and the volume of air sampled should be calculated or measured. There are four main types of denuder: cylindrical; annular; thermodenuders; and solution-based denuders. The recent development of thermodenuders is such that they have no use for solvent extraction as the adsorbed or absorbed species are heated off the inside of the denuder.

The diffusional loss of a gaseous analyte from an aerosol pumped through a denuder can be calculated by Equation 2.16, which was developed by Gormley and Kennedy (1949):

$$C/C_O = 0.819 \exp(-14.63\delta) \\ + 0.097 \exp(-89.2\delta) + 0.019 \exp(-212\delta) + \ldots \ldots \quad (2.16)$$

where, $\delta = \pi DL/4F$, C_O and C are the concentrations at the entrance and exit of the denuder, respectively, D is the diffusion coefficient of the

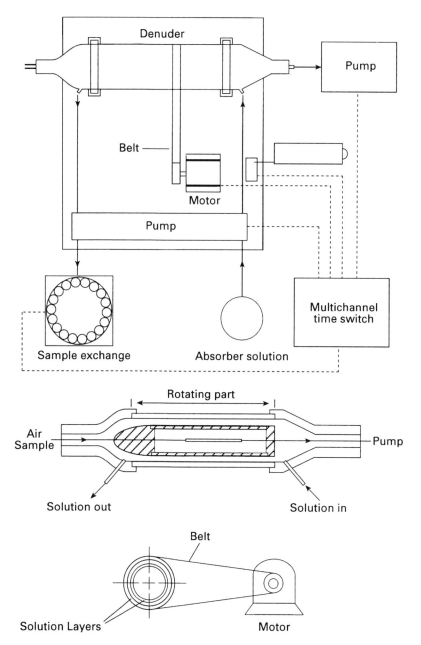

Figure 2.7 Schematic representation of a solution-based denuder. (Reproduced with permission from Nriagu (1992) *Gaseous Pollutants: Characterization and Cycling*, John Wiley and Sons, New York, USA, pp. 146–48).

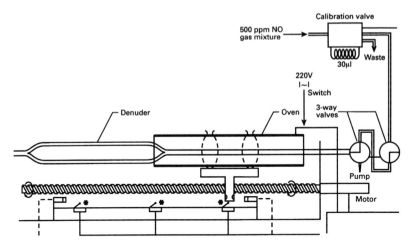

Figure 2.8 A schematic representation of a thermodenuder. (Reproduced from Nriagu (1992) *Gaseous Pollutants: Characterization and Cycling*, John Wiley and Sons, New York, USA, pp. 146–48).

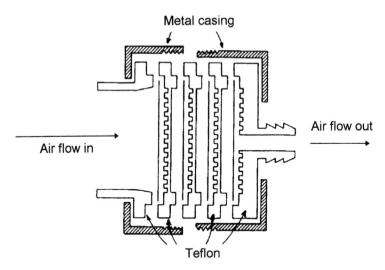

Figure 2.9 A holder for an impregnated filter showing the filters in position. (Reproduced with permission from Hewitt (1991) *Instrumental Analysis of Pollutants*, Elsevier Applied Science, London, UK, p. 5).

analyte collected in the aerosol (cm^2/s), L is the length of the denuder (cm), and F is the flow rate (cm^3/s). For an annular denuder (described in Section 2.3.2) higher flow rates are attained and the collection was determined by Possanzini *et al.* (1983) using Equation 2.17:

$$C/C_O = 0.819 \exp(-22.53\delta) \qquad (2.17)$$

where, $\delta = (\eta DL/4F)(d_1 + d_2/d_2 - d_1)$.

The flow rate is controlled by the size of the orifice. Equation 2.17 is only valid for laminar flow when the Reynolds number can be calculated using Equation 2.18, and it should be less than 2100:

$$\text{Re} = 4F/\eta vd \qquad (2.18)$$

where, v is the kinetic viscosity of air (0.152 cm^2/s stp) and d is the internal diameter of the denuder tube (cm). As laminar flow is not established until the aerosol has passed a critical distance within the tube, the entrance of the tube is not usually coated with the selective reagent. The distance to be left uncoated is calculated by Equation 2.19:

$$l = 0.07.\text{Re}.d \qquad (2.19)$$

Impregnated filters
This type of filter is used to extract gaseous analytes from a sample of air by passing the sample through a filter material that has been impregnated with a chemical reagent, which by chemical reaction will remove and concentrate the analyte of interest. Lindahl and co-workers (1996) described the theory of diffusive sampling within the impregnated filter. The theory is based on Fick's first law, which states:

$$m/t = DA(C - C_O)/L \qquad (2.20)$$

where, m is the mass collected on the filter (ng), t is the sampling time (s), D is the diffusion coefficient (cm^2/s), A is the cross-sectional area of the opening of the monitor (cm^2), C is the external concentration (ng/cm^3), C_O is the concentration of the analyte above the surface of the filter (ng/cm^3), and L is the length of the diffusive zone of the monitor (cm). If there is no reverse reaction at the filter, then the concentration of the analyte above the surface of the filter when the reaction is complete will be 0, i.e. when $C_O = 0$ then Equation 2.20 becomes:

$$m/t = DAC/L = SC \qquad (2.21)$$

where, S is the sampling rate, DA/L (cm^3/s). D is the diffusion coefficient and is analyte-dependent but A and L are parameters associated with the sampler equipment.

2.3.2 Instrumentation and methods

2.3.2.1 Denuders

(i) Cylindrical denuders. These have been applied to organic compounds as well as atmospheric constituents, and the instrumentation is shown in Figure 2.6; a long cylindrical tube with a selective coating.

(ii) Annular denuders. Annular denuder systems have been used to sample analytes, such as tetraethyllead (Febo *et al.*, 1986), lindane (γ- and α-hexachlorocyclohexane [γ-HCH and α-HCH]) (Johnson *et al.*, 1986) and C1-C3 aldehydes (Possanzini *et al.*, 1987). An annular denuder has a rod placed in the middle of the denuder tube (Figure 2.6b) that allows flow rates of up to 20–50 dm^3/min, which are ideal for the organic substances that can be present at very low concentrations.

(iii) Solution-based denuders. A schematic diagram of a solution-based denuder is presented in Figure 2.7. This denuder uses a microporous membrane surrounded by an outer tube. The aerosol is sucked through the main annulus or through the area between the membrane and the outer tube. The analyte of interest then diffuses toward the membrane and is collected in the selective solution which flows on one side of the microporous membrane.

(iv) Thermodenuders. The original thermodenuder involved the same apparatus as shown in Figure 2.8 but the particles from the aerosol were collected on the wall of the denuder, which was then heated. The heat caused the particulate material to decompose and the gases released then dissolved into the wall coating and the usual washing-off of the coating and off-line analysis followed. This process has been developed, such that when the particulate material decomposes the gases released are detected on-line. A typical system is shown in Figure 2.8.

2.3.2.2 Impregnated filters

Figure 2.9 shows the holder and set-up for an impregnated filter system. Sampling is performed by forcing air through a filter which is impregnated with a reagent with which the analyte of interest reacts. The simple impregnation of a filter with 2,4-dinitrophenylhydrazone (DPNH) for the determination of acetaldehyde in indoor air was described by Lindahl and co-workers (1996). Glass fibre filters were dipped into a solution containing DPNH, concentrated phosphoric acid, glycerol and acetonitrile, allowed to dry on a glass surface and subsequently placed in a filter holder.

2.3.3 Strengths and limitations

Oms-Molla and co-workers (1996) compared the sampling results for an impinger and a denuder. They reported that the denuder system produced much better recovery results, approximately 95%, whereas the impinger produced only 50% recovery due to loss of analyte on the walls of the inlet tube. They also suggested that one problem with the bubbler method was the large volume of solution required to dissolve the analyte, which then produced a low concentration for detection. Cylindrical denuders offer an easy, inexpensive and selective method for environmental analysis; however, the method is unattractive due to the long sampling times that are needed because of the low flow rates used (0.5–3 dm^3/min), and the labour-intensive extraction and second step off-line analysis required. These additional steps also increase the possibility of contamination. The problem of low flow rates is overcome by using an annular denuder, which uses much higher flow rates; consequently, enabling the sampling and analysis of organic compounds which are present in much smaller quantities. In intercomparison studies with similar techniques, annular denuders compare favourably and have similar accuracy and precision statistics. Whilst the advanced thermodenuders speed up the process of sampling by not requiring the coating to be washed-off and analysed off-line, there are problems with the stability of the coatings in the thermodenuders when higher temperatures are maintained. This is one of the problems which has caused the thermodenuder to compare unfavourably with other techniques.

Impregnated filters offer the following advantages: robustness and lightness of weight to transport when compared to solvent/bubbling extraction apparatus; the equipment occupies less volume; and because the filters are kept in housing, they are less prone to evaporation problems. Simultaneous collection of various analytes can be performed by using filters impregnated with different reagents in one housing system (Figure 2.9). Particulate matter can cause a problem with impregnated filters if no prefilter is used. Moreover, the capacity of an impregnated filter is limited and is reduced further if humidity is not controlled.

2.3.4 General areas of application

Aromatic hydrocarbons, such as benzene, toluene, m-xylene and o-xylene, were extracted from air samples using a charcoal-based denuder system (Risse *et al.*, 1996). Once the analytes were adsorbed onto the charcoal, they were extracted with either n-hexane or carbon disulphide by reclining the denuder tube at 30°C to the horizontal and rotating the tube slowly with the solvent in it. Gundel and co-workers (1995) described an annular denuder-based sampler, called the integrated organic vapour/particle sampler (IOVPS), which allows the direct determination both of gaseous and particulate semivolatile PAHs. A denuder-filter combination was used to sample and analyse gaseous and particulate nicotine (Häger and Niessner, 1996). These authors used a denuder coated with benzenesulphonic acid and a filter which had been impregnated with benzenesulphonic acid. The extracts were analysed by gas chromatography with a flame ionisation detector (FID). Peroxyacetyl nitrate (PAN) in ambient air has been measured using the nitrate ion and an alkaline carbon-coated denuder tube (De Santis *et al.*, 1996). Interferents were removed by placing two diffusion denuders infront of the annular denuder: the first coated with tetrachloromercurate (TCM) for the removal of sulphate and nitric acid; and the second coated with carbonate for the removal of HONO. Nicotine has been sampled in indoor environments using two annular denuders coated with benzenesulphonic acid and connected in series (Bertoni *et al.*, 1996). As an alternative to placing denuder tubes in series, Kallinger and Niessner (1997) have used a novel fourfold glass annular-denuder coated with acidified DNPH for the sampling of carbonyl compounds under high temperature and high humidity emission conditions, such as stack gases. After sampling, the hydrazones were eluted with acetonitrile, analyzed using reverse phase high-performance liquid chromatography (RP-HPLC) and detected via ultraviolet absorbance.

Lindahl and co-workers (1996) designed an impregnated filter system with a built-in control filter for determining acetaldehyde in indoor environments. In this system, half of the sampler was covered and so not exposed to the environment, which enabled this section to be used as a control or filter blank. Rudzinski and co-workers (1997) used an impinger system containing tryptamine to determine hexamethylene-based isocyanates in spray painting operations. They showed that the impinger system with dimethyl sulphoxide (DMSO) produced a better recovery than solid phase extraction procedures. In contrast, Wu and co-workers (1997) suggested that DMSO was too corrosive and highly hygroscopic for air sampling and showed that butyl acetate and octane were more suitable. Polyisocyanates are much sampled compounds, and Maitre and co-workers (1996) reported that impregnated filters significantly underestimated atmospheric concentrations of isocyanates. Spanne and co-workers (1996) determined complex mixtures of airborne isocyanates and amines using impingers containing dibutylamine (DBA) in toluene. The DBA derivatised isocyanates were then analysed using liquid chromatography with UV detection. When sampling PAHs in the workplace, Hiel and co-workers (1995) compared a stationary high-volume dust sampling system equipped with a glass fibre filter and a stationary sampling device consisting of a glass wool filter and serially connected impingers. They found that the dust collection system underestimated the total PAH concentration, although the filter impinger system was considered too complicated for routine use at heavy duty worksites. Oms-Molla and Klockow (1995) found that the collection efficiency of denuders was much higher than that of an impinger system.

Maitre and co-workers (1996) found that 1-(2-methoxyphenyl)piperazine (MPP)-impregnated filters constantly underestimated the exposure to atmospheric 1,6-hexamethylene diisocyanate (HDI) polyisocyanates when compared with the use of an impinger system using 1-(2-methoxyphenyl)piperazine (MPP) as the absorbing solution. Cyclic organic acid anhydrides (OAAs), which are used in alkyd and epoxy resins and in the production of plasticizers, are known irritants to the eyes and to the mucous membranes of the respiratory tract and have been shown to produce allergic rhinitis and asthma. Because of these properties, there is a need to monitor OAAs in the workplace. Jönsson and co-workers (1996) reviewed the monitoring of OAAs using solid sorbents, bubblers and impingers. There is a protocol by which a diffusive sampler can be tested, which has been adopted by the European Committee for Standardisation (CEN) (1995). The protocol describes the test to be performed to determine the effect that sampling time, concentration, relative humidity, temperature, storage and wind velocity have on the sampling rate. Takeshita and co-workers (1995) used an impinger system to enhance the sorption of polychlorinated dibenzofurans (PCDD/F) in flue gases onto DS-sephadex column; a solid sorbent. The impinger system which contained PR-DEAE-sephadex, cellulose powder and water was used to remove acidic substances which would inhibit the sorption of PCDD/Fs onto the solid sorbent.

2.4 Membrane extraction

A membrane system is defined as two essentially uniform and homogeneous three-dimensional fluid phases, between which matter and energy may be exchanged at rates governed by the properties of a third phase, or group of phases, which separates them. The third phase is known as the membrane (Meares, 1986). A membrane is defined by Hampel and Hawley (1976) as an extremely thin, porous layer or film of material, either natural or synthetic, with an approximate thickness of 10 nm and with the diameter of the pores ranging from as little as 0.2 µm for natural cellular membranes up to 10 nm for manufactured membranes. Membranes which have pores so small that only the molecules of solvents, ions and a few substances of very low molecular weight can pass through are described as semipermeable. Within this section, membrane extraction has been taken to be synonymous with membrane filtration. Membrane filtration methods can be referred to as either macrofiltration, ultrafiltration, nanofiltration or hyperfiltration; which filtration method is used is defined by the pore size of the membrane. Table 2.7 gives details of some of the membrane extraction systems that are available.

2.4.1 Basic principles

The basic principles of how heterogeneous membranes separate selective solutes has been described in detail by Meares (1986), and so only the more simple homogeneous situation will be described in this chapter.

As stated previously, Fick's first and second laws underpin much of solvent and membrane extraction theory. In a homogeneous membrane extraction system, substances can enter the membrane by dissolving into the membrane material. The separation process relies on this step being thermodynamically stable. The concentration of analyte which dissolves into the membrane is governed by the concentration of the analyte in the initial sample, assuming that: (i) the concentrations of the analyte in the phases external to the membrane are uniform; (ii) the partition equilibrium of the filtrate is maintained at the external phases, such that the free energy barriers opposing the entry and release of the analyte are negligible; (iii) the partition coefficient (k) of the filtrate between the membrane and the adjacent phases obeys Henry's law and is independent of the concentration of the analyte in the initial sample; and (iv) the diffusion coefficient (D) of the filtrate is independent of its concentration (C) in the membrane; such that then, under ideal conditions, the flux (flow rate) for the analyte is given by Equation 2.22.

$$J = P(C_{OI} - C_{OE})/l \qquad (2.22)$$

where, J is the steady flux density (mol/m²/s), l is the membrane thickness, P is the permeability coefficient of the filtrate in the membrane at the specified temperature; and that C_{OI} is the concentration of the analyte in the initial sample and it is extracted into a second phase (the membrane) at the concentration C_{OE}. This theory is relevant whether the analyte is a gas or a liquid.

In ultrafiltration, the following equation can be used to determine whether a membrane will reject a solute from a sample:

$$R = \downarrow n(C_f/C_o)/ \downarrow n(V_o/V_f) \qquad (2.23)$$

where, C_o and V_o are the initial solute concentration in a sample and the initial sample volume, respectively, and C_f and V_f are the final solute concentration in a sample and the retentate volume, respectively. When the solute is not extracted by the membrane, then $R = 1$. A process occurs

Table 2.7 Some of the membrane extraction processes available

Process	Driving force (gradient)	Permeating species	Species retained
Microfiltration (MF)	Pressure 100–500 kPa	Solvent (water) and dissolved solutes	Suspended solids, fine particles and some colloids (1,000,000–100,000,000)
Ultrafiltration (UF)	Pressure 100–800 kPa	Solvent (water), salts, lower MW macromolecules	Higher MW macromolecules and colloids (500–1,000,000)
Nanofiltration (NF)	Pressure 0.3–3 MPa	Solvent (water), low MW compounds and univalent ions	High MW compounds (>100–1000) and multivalent ions
Reverse osmosis (RO)	Pressure 1–10 MPa	Solvent (water)	Virtually all dissolved solids and suspended solids
Electrodialysis (ED)	Electric potential 1–2 V/cell pair	Solutes (ions), small quantity of solvent	Non-ionic and macromolecular species
Dialysis (D)	Concentration	Solute (ions and low MW organics) and small amounts of solvent	Dissolved and suspended solids with MW > 1000
Gas permeation (GS)	Pressure 0.1–10 MPa	Highly permeable gases and vapours	Less permeable gases and vapours
Pervaporation (PV)	Chemical potential or concentration	Highly permeable solute or solvent	Less permeable solvent or solute
Supported liquid membrane (SLM)	Concentration	Ions, low MW organics or gases	Ions, less permeable organics or gases
Membrane distillation (MD)	Temperature	Volatile species (water vapour)	Nonvolatile species
Pertraction (PT)	Chemical potential or concentration	Highly permeable gases, solutes or solvent	Less permeable gases, solutes or solvents

Abbreviations: MW, molecular weight. (Modified from Golemme and Drioli, 1996).

during membrane extraction called concentration polarization, when there is a build-up of unfiltered solute on the surface of the membrane. This is when the solute cannot pass through the membrane. In time, the effect of concentration polarization creates a gel-type layer on the surface of the membrane, which acts as a second membrane surface. The flow rate of the solute through the new biomembrane formation is defined by Equation 2.24:

$$J = (\Delta P - \Delta \pi)/(R_g + R_m) \quad (2.24)$$

where, ΔP is the transmembrane pressure drop, $\Delta \pi$ is the osmotic pressure of the solution, R_g is the hydraulic resistance of the gel layer, and R_m is the hydraulic resistance of the membrane. As the osmotic pressure for the macrosolutes in solution and colloidal dispersions is very low, Equation 2.24 can be simplified to:

$$J = \Delta P/(R_g + R_m) \quad (2.25)$$

At high solute concentrations, the solute precipitates at the membrane surface and begins to form the gel-type layer, which allows solvent only to pass through. At lower concentrations of the solute, the resistance of the gel layer is more significant than the membrane itself and the flow rate becomes independent of the membrane pore size. As this process continues and the pressure is continually applied, the solute continues to accumulate at the membrane/gel interface, thus slowing the flux of the solute. Resistance continues to increase until the net transport of the solute due to solvent flow equals the back diffusion of the solute towards the bulk solution, due to the concentration gradient. Even if further pressure is applied at this point, the flow rate will remain the same and the gel layer will thicken further. When the concentration of the solute in the filtrate is very low, the steady-state of flux of solvent through the membrane is expressed as:

$$J(C) + D_s(dc/dy) = 0 \quad (2.26)$$

where, D_s is the diffusivity of the solute. The final equation for this system is:

$$J = K \downarrow n(C_g/C_s) \quad (2.27)$$

where, C_g is the concentration of the gel, and C_s is the bulk concentration of the solution. K is the mass-transfer coefficient, which is a measure of the flow of the solute away from the membrane surface and is also affected by temperature. For example, for proteins, the solute diffusivity increases by 3–3.5% per °C.

In solvent extraction the choice of the extracting solvent is pivotal and in membrane extraction the choice of the membrane is of great importance. Meares (1986) described four basic types of transport medium for membranes: (i) Permanent pores (>5 nm diameter) intentionally introduced into the polymeric membrane during manufacture. Transport of small molecules through these pores is primarily via convection and with only little selectivity. (ii) Small pores (<1 nm diameter), which are formed naturally in some solid state polymers and which can be interconnecting and continuous. The flow of solutes will be via convection and some diffusion. (iii) Amorphous polymers, which are liquid-like on the local scale and the solutes can dissolve into the polymer and be transported via concentration gradients. (iv) Gel substances, which involve a polymer mixing with a liquid to create a gel substance through which solutes are transported via diffusion and convection. The characteristics of these membrane systems are summarised in Table 2.8.

(i) Ultrafiltration
Membranes can exist in various forms: tubular; spiral, wound; multi-channel monolith elements; hollow fibre; and flat sheet. Each membrane has an associated 'cut-off' value. This value should only be taken as an indicator, as it will reject 90% of the material at the size of the cut-off value; thus, it is recommended that, when wanting to separate molecules of a specific size, a membrane with a cut-off value well below that of the solute to be retained is selected. The cut-off value and, hence, pore-size of the membrane will determine the flow rate attainable with the membrane. As ultrafiltration is a pressure-driven membrane system, the pore diameters are usually in the range 5–50 nm and so molecules in the molecular weight range 300–500,000 are rejected (Koops, 1995). Other factors which can affect the flow rate are: surface area of membrane; macrosolute type; solubility; concentration of solute and diffusivity; membrane type and flow channel dimensions and characteristics; temperature; and, to some extent, pressure. Golemme and Drioli (1996) suggested that the important membrane characteristics for ultrafiltration are: porosity; morphology; resistance; surface properties; mechanical strength; chemical resistance; thermal resistance; and compactness.

(ii) Nanofiltration
Again, nanofiltration is a pressure-driven process, where components are extracted according to their size (molecular weights usually 100–1000) and charge. The pore diameter in a nanofiltration membrane is usually 1–3 nm (Koops, 1995).

Table 2.8 Characteristics of the four different membrane systems

Membrane type	Pore size	Pore details	Substances retained	Transport mechanisms	Uses
Macroporous	Irregular, wide range	Interconnecting network	Macromolecules, colloidal materials	Convective flow is rather greater than diffusion or conduction	Macromolecular sieve, dialysis and ultrafiltration of colloids, semipermeable membranes for polymer separation
Microporous	Diameter comparable with thickness of polymer chains	Irregular in position, changes with temperature	Bacteria, colloids, particulates > 0.1 μm diameter	Combination of convection and diffusion	Gas and liquid vapour separation
Solvent-type	Not defined, irregular	Substances dissolve into the polymer and migrate down the concentration gradient	Low moelcular weight substances	Via concentration gradient, controlled by the random thermal motions of the solute polymer mixture	Extract analytes from vapours
Gel	< 40–300 μm			Diffusion and convection	

(iii) Gel extraction
In gel filtration/extraction, the polymeric material has formed pores within the gel matrix. There are a wide variety of gel materials available and the one to choose is dependent upon the size of the solutes to be separated. In a gel-solution system, small molecules will diffuse into the gel material from the initial solution; larger molecules will not diffuse into the gel to the same degree as the small molecules because of their molecular size. Very large molecules (compared to the pore size of the gel) will stay in the initial solution that the analyte is contained in. Again, the elution or extraction of an analyte using gel filtration can be described by the distribution coefficient, K_D, which represents the fraction of the gel which is available for diffusion of a given analyte. Similarly, K_{av} defines solute behaviour independently of the bed dimensions and packing. K_{av} can be calculated as follows:

$$K_{av} = (V_e - V_o)/(V_t - V_o) \quad (2.28)$$

where, V_e is the elution volume, V_o is the void volume, and V_t is the total volume of the packed bed.

2.4.2 Instrumentation and methods

Table 2.7 shows the various membrane extraction processes available. Membrane materials are numerous; various materials are described in Table 2.9. Macroporous membranes are prepared by applying a solvent and a nonsolvent to a polymeric material, either sequentially or simultaneously, to create the irregular pores. With microporous membranes, a solvent is applied to irregularly packed films of amorphous polymers. A cross-section of a microporous membrane is shown in Figure 2.10. Solvent-type membranes are submerged elastomeric and flexible polymers in solvent materials, and gel membranes are prepared by swelling a polymeric material with a liquid. The degree of swelling is dependent upon the volume of the liquid used to swell the material.

(i) Ultrafiltration, nanofiltration and hyperfiltration
In ultrafiltration, depending upon the pore size of the extraction membrane, membranes can be classified as macroporous or microporous. Ultrafiltration membranes have two components. The first is a thin extraction membrane (5–50 nm) which has a precisely controlled pore size. The second part of the membrane has an open structure, starting with small pores next to the extraction membrane, which then get larger and develop into voids the further away from the extraction membrane the sample filtrate progresses. Laboratory ultrafiltration apparatus has a

Table 2.9 Membrane materials and their applications

Type	Properties	Application	Comment
Cellulose acetate	Minimal protein binding	Aqueous samples; protein recovery	Used in conjunction with a glass prefilter can filter tissue cultures and sensitive biological samples
Mixed cellulose esters	Mixture of acetate and nitrate; hydrophilic; autoclavable to 121°C	Biologically inert; clarification of fluids; sterilizing filtration of tissue culture fluids; air monitoring	
Glass media	Borosilicate microfibre glass with or without acrylic binder resin	DNA recovery and clean up; prefilter Without binder; determination of volatile suspended matter in waste water, industrial effluents and aerosol monitoring	Used as prefilters
Nitrocellulose	Hydrophilic; limited solvent resistance; high protein binding		DNA and RNA blotting
Polycarbonate film	Hydrophilic; autoclavable to 121°C	Beverage testing; sterility testing; bioassays; air analysis	Filter is translucent with a slight green tinge
Nylon	Solvent resistant; binds proteins; autoclavable; withstands gamma radiation or ethylene oxide; withstands temperatures between -40 and $+115°C$	Particulate filtration; paint monitoring	
Polyethersulphone	Hydrophobic membrane, high flow characteristics; moderate protein binding; fairly solvent resistant		
Polypropylene	Hydrophilic membrane; compatible with organic solvents; withstands temperatures up to 90°C	Biological sample filtration; protein recovery	

Material	Properties	Uses	Comments
PTFE	Hydrophobic; chemically resistant to solvents, acids, bases	Degassing chromatography solvents; clarification of acids, bases and alcohol solutions	Membrane can be rendered hydrophilic if wetted with methanol or other low surface tension fluid; available without backing to allow extreme temperatures and chemical conditions
PVC	Medical grade PVC	Air monitoring of silica, carbon black and quartz particulates	
Hydrophilized PTFE		Hydraulic fluids, RNA isolation; clarification of propellants	
PVDF (durapore)	Hydrophilic and hydrophobic; solvent resistant; low resistance to UV absorbing extractables; low protein binding	Hydrophilic has a low binding capacity and hydrophobic has a high binding capacity	Hydrophilic durapore is modified PVDF and hydrophobic durapore is PVDF
Regenerated cellulose	Hydrophilic; solvent resistant; low protein binding; autoclavable; withstands gamma radiation or ethylene oxide	Biological samples; maximum recovery of protein	
Ultrafiltration	Pore size range 0.0001–0.01 μm	Separation of various sized compounds, 5000–100,000 Da; protein concentration; salt removal	Centrifugal ultrafilters available

Abbreviations: PTFE, polytetrafluoroethane; PVC, polyvinyl chloride; PVDF, polyvinylidene fluoride; DNA, deoxyribonucleic acid; RNA, ribonucleic acid.

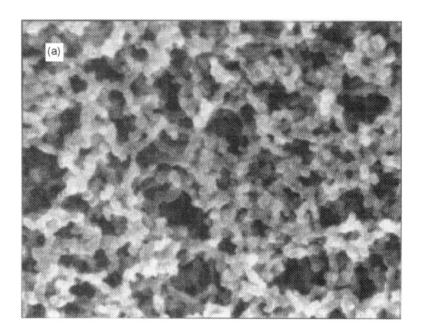

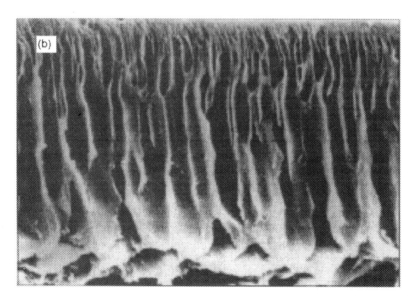

Figure 2.10 A microporous membrane. (Reproduced from Amicon catalogue, publication No. 277E, with permission from Millipore, UK, Ltd).

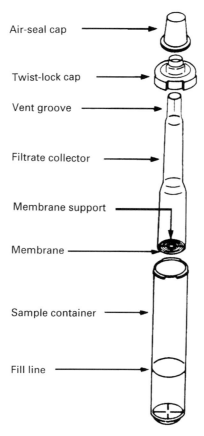

Figure 2.11 The components of an ultrafiltration Centriprep® concentrator system. (Reproduced with permission from Millipore, UK, Ltd).

generic operation. The sample is pipetted into a container which has the membrane at its bottom, this container is then placed on top of or inside a second container. The entire system is then placed in a centrifuge and spun. The components of an ultrafiltration centrifuge system are shown in Figure 2.11 and the process is explained in Figure 2.12. The difference between nanofiltration and hyperfiltration is the pore size of the membrane used. Membrane hyperfiltration allows the separation of species with molecular masses of 100–500. Currently, when membranes are used for either ultrafiltration, nanofiltration or hyperfiltration, they are housed in modified acrylic, polypropylene or stainless steel casings. The materials from which the analyte(s) are to be extracted are either pumped or sucked through the filter, via pressure or vacuum pumps.

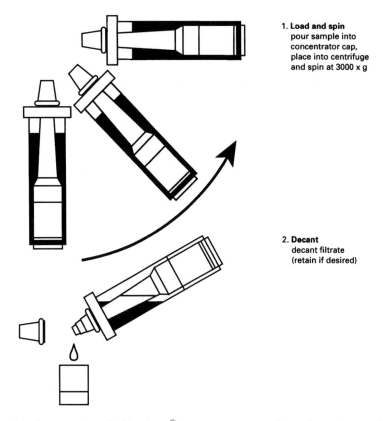

Figure 2.12 The operation of a Centriprep® concentrator system. (Reproduced with permission from Millipore, UK, Ltd).

(ii) Gel extraction
Generally, gel extraction is most conveniently achieved by placing the gel material in a tubular column. The columns can be packed manually or bought prepacked from suppliers. Extraction of the analyte of interest is achieved by placing the initial sample containing the analyte onto the top of a column which has been packed with the required gel material. The sample is then pushed down the column with a solvent system, which will not interact with the sample using either external pressure or gravity. The eluant is collected according to when the analyte of interest is likely to elute. This can be established using molecular markers of known size and recording the volume of eluant that is required to elute the various sized markers.

(iii) Pervaporation
Pervaporation is extraction by partial vaporization (Favre *et al.*, 1996) and is described mainly in the literature associated with methods in biotechnology in the improvement of fermentation products, etc. It is a relatively new separation process that uses a dense polymeric membrane, above which there is a liquid mixture and below which a vacuum is created. The analyte of interest undergoes a phase change (from liquid to gas) due to a change in the partial pressure of the permeate, which is achieved by using the vacuum pump. The features of pervaporation include: (i) no membrane stability problems; (ii) no fouling of membranes; (iii) heat from exothermic bioreactions can be released in the pervaporation unit; (iv) no thermal, chemical or mechanical stress on the fermentation broth; and (v) can increase productivity (Favre *et al.*, 1996).

Analyte transport in pervaporation is based on solution-diffusion mechanisms. The transport system is composed of two steps: sorption equilibrium at the upstream part of the membrane (S); and the diffusion of the permeants through the polymer network (D). The product of the two components is permeability (P). Therefore, the pervaporation flux is the result of the permeability (P) and the concentration gradient created:

$$J = (DS\Delta c/z) = (P\Delta c/z) \qquad (2.29)$$

2.4.3 Strengths and limitations

Ultrafiltration is a much gentler process than nonmembrane techniques and, unlike other separation methods, it does not require a phase change which can denature labile species. Care must be taken if temperature is to be increased when using ultrafiltration with fermentation broths in the presence of anti-foams. A point called 'cloud point' is possible, which occurs when the temperature has increased too far and has caused the anti-foam to come out of solution and deposit on the membrane, creating a second artificial layer. Users of ultrafiltration need to be aware that molecules of similar size may exhibit different filtering capacities due to their degree of hydration, counter ions and stearic hindrances. Moreover, biological molecules have a tendency to aggregate and/or change their conformation due to the processes of membrane extraction, altering conditions such as pH and ionic strength.

Many workers report the effects of flux reduction on the efficiency of membrane separations. Jönsson and co-workers (1997) demonstrated that the reduction in flux was due to the retention of solutes in the flow channels of the membrane and not caused by the solutes forming a cake or gel layer at the surface of the membrane. In nanofiltration, Hong and Elimelech (1997) demonstrated that membrane fouling increased with increasing electrolyte concentration, decreasing solution pH and the addition of divalent cations, such as calcium. They also reported that the rate of fouling was controlled by a combination of permeation drag and electrostatic double-layer repulsion.

Nanofiltration has been used to separate eluent components of bleaching pulp and paper (Afonso and de Pinho, 1997). The nanofiltration membrane consisted of an amine and an ampholytic polymer containing quaternary amine groups and sulphonic acid groups. The efficiency of the membrane was determined by the ampholytic polymer, as it increased the membrane hydrophilicity causing high permeation fluxes and a low fouling tendency.

Gustafsson and co-workers (1996) discussed the use of crossflow filtration (CFF) for sampling marine organic colloids, and the need to condition the CFF membrane immediately before sampling to reduce the effect of contamination. The advantage of pervaporation as a membrane extraction method is that it allows the extraction of temperature sensitive volatiles, such as dairy aroma compounds, as reported and discussed by Baudot and Marin (1996).

2.4.4 General areas of application

One of the most obvious applications of ultrafiltration is dialysis. Dialysis is a technique which uses the principle that different compounds will diffuse through a membrane at different rates; a typical use of dialysis is to isolate organic compounds from biological samples, e.g. blood plasma or urine. The equipment used for dialysis involves a bag-shaped membrane (which contains the sample from which an analyte is to be extracted, e.g. organic analyte in a water sample) suspended in water. The low molecular weight material will then move from a situation where it is in high concentration to one where it is more dilute, and so the small molecules will move into the surrounding water through the membrane. The rate of diffusion can be increased by stirring or by replacing the surrounding water periodically (comparable to multiple solvent extraction with a separating funnel). A development of dialysis is the more rapid form, electrodialysis, where electrolyte solutes are removed from neutral, organic material. A difference in potential is applied at points on either side of a cell containing the solution to be dialysed. The voltage applied increases the rate of diffusion. This method could be applied to samples to remove unwanted impurities which may hinder the final analysis (Pinder, 1964).

The solubility of membrane filters in a solvent has been exploited to determine trace amounts of aniline in seawater samples (Gu et al., 1997). A standard spectrophotometric method was used; in a medium of 0.1 M hydrochloric acid, aniline diazotizes with nitrous acid, which is then coupled with N-(1-naphthyl)-ethylenediamine. The azo-dye that is formed was collected onto a 0.45 μm nitrocellulose filter maintained at a pH of 5 in the presence of tetradecylpyridinium (TDCP) surfactant. When sufficient washings had been performed to reduce the reagent blank, the filter containing the azo-dye compound was dissolved in a volume of methoxyethanol acidified with hydrochloric acid. The resulting solution was measured at 560 nm. The detection limits for aniline were better than $1.0\,\mu g/dm^3$. A similar adaptation of membrane filtration was described by Kaul and Mattiasson (1992) for the separation of proteins. They covalently attached an affinity ligand to an ultrafiltration membrane. This allowed the retention of the protein, because it bound to the ligand whilst the unbound impurities passed through the filter. After the required washing, the bound protein could be desorbed by passing a dissociating medium through the filter; the selected protein was then collected in the retentate. The membrane with the bound ligand was reinitialised by washings to remove the dissociating medium, after which it was ready to use. The membrane affinity concept combines the clarification, purification and concentration steps. This is possibly similar to the use of gels as an extraction mode.

Brose and co-workers (1995) described a method which used a membrane-based extraction process for the fractionation of citrus oils to produce an oil stream of enriched oxygenated components. The procedure involved flowing citrus oil on one side of a nonporous membrane with an aqueous cyclodextrin solution on the other. The oxygenates diffused through the water-swollen hydrophilic membrane into the aqueous cyclodextrin solution. The cyclodextrin-oxygenate complex was then dissociated by passing the complex through a second membrane, causing the oxygenates to diffuse into a citrus oil solution. The use of flat-sheet supported liquid membranes was compared with hollow-fibre membrane extraction systems for the removal of organic acids from propionic and acetic acid fermentations (Ozadali et al., 1996). These authors concluded that the limit on the acid production was attributable to the substrate feed and not the efficiency of the extraction system.

A variation on the filtration theme is crossflow filtration, in which the main retentate flow is parallel to the filter surface and, after penetrating the depth of the filter membrane, the permeate will flow along the filter plane (Gustafsson et al., 1996). CFF has allowed the use of high flux pressures with small pore sizes, thereby reducing the effects of concentration polarization. Despite this development, the method still suffers from polarization and fouling effects.

2.5 Overall future directions and developments

At the present time, the future application of solvent extraction has manifested itself in the development of techniques such as solid phase extraction, supercritical fluid extraction, accelerated/enhanced solvent extraction and microwave extraction. The exact developments of these techniques are fully described in Chapters 3, 5, 6 and 7, respectively. The inevitable combination of techniques in solid phase membrane extraction (SPME) and membrane extraction with sorbent interface (MESI) are described in Chapter 4.

The success of membrane separation is dependent upon the material of the membrane as well as the design, and hence the future development of membrane extraction appears to be in these areas. Research is concentrating on the development of new materials which can withstand extreme conditions of temperature, pressure and corrosivity. Golemme and Drioli (1996) reviewed the use of poly(organophosphazene)s (POPs) as a membrane material for ultrafiltration, nanofiltration, pervaporation, vapour permeation and gas separation. They have suggested that, by combining POP with inorganic materials, the gap between polymer and ceramic membranes may be filled. Genné and co-workers (1997) reported on a new membrane material, which combined the flexibility properties of organic membranes with the properties of inorganic membranes, such as pressure resistance and surface properties. The organo-mineral combination used by these authors was polysulphone and zirconia. Their paper illustrates the potential to increase the permeability of a membrane and its surface properties by adding suitable amounts of inorganic fillers to an organic base.

A number of articles have reported on the use of supported liquid membranes (Wieczorek et al., 1997; Harriott and Ho, 1997; Munro and Smith, 1997; Juang et al., 1997a; Juang et al., 1997b). Wieczorek and co-workers (1997) used supported liquid membranes to concentrate amino acids. The liquid membrane contained di-2-ethylhexyl phosphoric acid, and the amino acids were extracted from an aqueous donor phase to a more acidic acceptor phase. The movement was governed by the proton gradient. An extraction efficiency of 60% was obtained for approximately 12 h, which produced an enrichment factor of 150.

Other workers are attempting to couple the sample preparation step of membrane extraction with the analytical technique. To reduce the need and time factor of lengthy sample preparation, Garcia Sanchez and co-workers (1997) have coupled a membrane extraction system with a high-performance liquid chromatographic (HPLC) system with electrochemical detection for the extraction of phenols from crude oils. A silicone membrane was used (Garcia Sanchez *et al.*, 1997).

In addition to supported liquid membranes, there are emulsions which can be used for liquid membrane extractions. Emulsion liquid membranes (ELMs) involve the extraction of the analyte in an aqueous solution into a multicomponent emulsion. They are first prepared by forming an emulsion between two immiscible phases, followed by the dispersion of this emulsion into a third phase by agitation. The membrane is the liquid which separates the internal micelles of the emulsion from the third liquid phase (Yurtov and Koroleva, 1996).

Gustafsson and co-workers (1996) suggest that future work with crossflow filtration should concentrate on the optimisation of the instrumental design. Initially, this procedure would be aided if workers, when reporting their findings, describe the testing and calibration of their systems with standard colloid materials. A sensible and timely request, considering the upsurge in valid analytical measurement that is taking place.

References

Afonso, M.D. and de Pinho, M.N. (1997) Nanofiltration of bleaching pulp and paper effluents in tubular polymeric membranes. *Sep. Sci. Technol.*, **32** 2641-58.
Ali, Z., Thomas, C.L.P. and Alder, J.F. (1989) Denuder tubes for sampling gaseous species. *Analyst*, **114** 759-69.
Atkins, P. (1986) *Physical Chemistry*, 3rd edn, Oxford University Press, Oxford, UK.
Baudot, A. and Marin, M. (1996) Dairy aroma compounds recovery by pervaporation. *J. Membr. Sci.*, **120** 207-20.
Bertoni, G., Di Palo, V., Tappa, R. and Possanzini, M. (1996) Fast determination of nicotine and 3-ethylpyridine in indoor environments. *Chromatographia*, **43** 296-300.
Braghetta, A., DiGiano, F.A. and Ball, W.P. (1997) Nanofiltration of natural organic matter: pH and ionic strength effects. *J. Environ. Engin.*, **7** 628-41.
Brose, D.J., Chidlaw, M.B., Friessen, D.T., Lachapelle, E.D. and Van Eikeren, P. (1995) Fractionation of citrus oils using a membrane-based extraction process. *Biotechnology Progress*, **11** 214-20.
Coutant, R.W., Callhan, P.J., Chuang, J.C. and Lewis, R.G. (1992) Efficiency of silicone-grease-coated denuders for collection of polynuclear aromatic hydrocarbons. *Atmos. Environ.*, **9** 517-22.
De Santis, F., Allegrini, I., Di Filippo, P. and Pasella, D. (1996) Simultaneous determination of nitrogen dioxide and peroxyacetyl nitrate in ambient atmosphere by carbon-coated annular diffusion denuder. *Atmos. Environ.*, **30** 2637-45.
European Committee for Standardisation (CEN) (1995) Workplace Atmospheres—Diffusive Samplers for the Determination of Gases and Vapours. Requirements and Test Methods. EN 838: 1995, CEN, Brussels, Belgium.

Falandysz, J., Florek, A., Kulp, S., Stranberg, B., Strandberg, L., Bergqvist, P. and Rappe, C. (1996) *Bromat. Chem. Toksykol*, **XXIX** 267-70.

Favre, E., Nguyen, Q.T. and Bruneau, S. (1996) Extraction of 1-butanol from aqueous solutions by pervaporation. *J. Chem. Tech. Biotechnol.*, **65** 221-28.

Febo, A., di Palo, V. and Possanzini, M. (1986) The determination of tetra-alkyl lead in air by a denuder diffusion technique. *Sci. Total Environ.*, **48** 187-94.

Garcia Sanchez, M.T., Perez Pavon, J.L. and Moreno Cordero, B. (1997) Continuous membrane extraction of phenols from crude oils followed by high performance liquid chromatographic determination with electrochemical detection. *J. Chromatogr. A*, **766** 61-69.

Genné, I., Doyen, W., Adriansens, W. and Leysen, R. (1997) Organomineral ultrafiltration membranes. *Filtration & Separation*, November, 964-66.

Golemme, G. and Drioli, E. (1996) Polyphosphazene membrane separations (Review). *J. Inorg. Organomet. Polymers*, **6** 341-65.

Gormley, P.G. and Kennedy, M. (1949) Diffusion from a stream flowing through a cylindrical tube. *Proc. R. Ir. Acad. Sect. A*, **52** 103.

Grinshpun, S.A., Willeke, K., Ulevicius, V., Juozaitis, A., Terzieva, S., Donnelly, J., Stelma, G.N. and Brenner, K.P. (1997) Effect of impaction, bounce and re-aerosolization on the collection efficiency of impingers. *Aerosol Sci. and Technol.*, **26** 326-42.

Grömping, A.H.J. and Cammann, K. (1996) Field evaluation and automation of a method for the simultaneous determination of nitrogen oxides, aldehydes and ketones in air. *J. Automatic Chemistry*, **18** 121-26.

Gu, X., Li, C., Qi, X. and Zhou, T. (1997) Determination of trace aniline in water by a spectrophotometric method after preconcentration on an organic solvent-soluble membrane filter. *Anal. Letts*, **30** 259-70.

Guillén, M.D. (1994) Polycyclic aromatic compounds: extraction and determination in food. *Food Adds. Contam.*, **11** 669-84.

Gundel, L.A., Lee, V.C., Mahanama, K.R.R., Stevens, R. and Daisey, J.M. (1995) Direct determination of the phase distributions of semi-volatile polycylic aromatic hydrocarbons using annular denuders. *Atmos. Environ.*, **29** 1719-33.

Gustafsson, O., Buessler, K.O. and Gshwend, P.M. (1996) On the integrity of crossflow filtration for collecting marine organic colloids. *Mar. Chem.*, **55** 93-111.

Häger, B. and Niessner, R. (1996) On the distribution of nicotine between the gas and particle phase and its measurement. *Aerosol Sci. and Technol.*, **26** 163-74.

Hampel, C.A. and Hawley, G.G. (1976) *Glossary of Chemical Terms*, Van Nostrand Reinhold Co., New York, p. 172.

Hancock, P. and Dean, J.R. (1997) Extraction and fate of phenols in soil. *Anal. Comm.*, **34** 377-79.

Harriott, P. and Ho, S.V. (1997) Mass transfer analysis of extraction with a supported polymeric liquid membrane. *J. Membr. Sci.*, **135** 55-63.

Hewit, C.N. (1991) *Instrumental Analysis of Pollutants*. Elsevier Applied Science, London, UK, p. 5.

Heyward, M.P., Hurle, R.L. and Sauerhammer, B. (1997) Determination of the recovery of dimethylmercury and diphenylmercury extracted from organic solvents and a liquid condensate with bromine water using cold vapour atomic absorption spectrometry. *Anal. Comm.*, **34** 279-81.

Hiel, N., Nyiry, W. and Winker, N. (1995) High volume sampling polycyclic aromatic hydrocarbons in the workplace. *Fres. J. Anal. Chem.*, **352** 725-29.

Hong, S. and Elimelech, M. (1997) Chemical and physical aspects of natural organic matter (NOM) fouling of nanofiltration membranes. *J. Membr. Sci.*, **132** 159-81.

Hyötyläinen, J., Knuutinen, J., Malkavaara, P. and Siltala, J. (1998) Pyrolysis-GC-MS and CuO-oxidation-HPLC in the characterization of HMMs from sediments and surface waters downstream of a pulp mill. *Chemosphere*, **36** 297-314.

Isaacs, A., Daintith, J. and Martin, E. (eds.) (1989) *Concise Science Dictionary*, Oxford University Press, Oxford, p. 649.
Johnson, N.D., Barton, S.C., Thomas, G.H. S., Lane, D.A. and Schroeder, W.H. (1986) Field evaluation of a diffusion denuder-based gas/particle sampler for chlorinated organic compounds. 79th Annual Meeting of the Air Pollution Control Association, Minneapolis, MN, USA, June 22-27, 1986.
Jönsson, A.-S., Lindau, J., Wimmerstedt, R., Brinck, J. and Jönsson, B. (1997) Influence of the concentration of a low-molecular weight organic solute on the flux reduction of a polyethersulphone ultrafiltration membrane. *J. Membr. Sci.*, **135** 117-28.
Jönsson, B.A.G., Welinder, H. and Pfäffli, P. (1996) Determination of cyclic organic acid anhydrides in air using gas chromatography. Part 1. A review. *Analyst*, **121** 1279-84.
Juang, R.S., Huang, R.H. and Wu, R.T. (1997a) Separation of citric and lactic acids in aqueous solutions by solvent extraction and liquid membrane processes. *J. Membr. Sci.*, **136** 89-99.
Juang, R.S., Lee, S.H. and Shiau, R.C. (1997b) Mass-transfer modelling of permeation of lactic acid across amine-mediated supported liquid membranes. *J. Membr. Sci.*, **137** 231-39.
Kallinger, G. and Niessner, R. (1997) Development and laboratory investigation of a denuder sampling system for the determination of carbonyl compounds in stack gas. *Fres. J. Anal. Chem.*, **358** 687-93.
Kaul, R. and Mattiasson, B. (1992) Secondary purification. *Bioseparations*, **3** 1-26.
Kelsey, J., Kottler, B.D. and Alexander, M. (1997) Selective chemical extractants to predict bioavailability of soil-aged organic chemicals. *Environ. Sci. Technol.*, **31** 214-17.
Koops, G.H. (1995) *Nomenclature and Symbols in Membrane Science and Technology*. European Society of Membrane Science and Technology, Toulouse and Enschede, p. 23.
Lewis, R.J. (1993) *Hawley's Condensed Chemical Dictionary*, 12th edn, Van Nostrand Reinhold, New York, USA.
Lin, X., Willeke, K., Ulevicius, V. and Grinshpun, S.A. (1997) Effect of sampling time on the collection efficiency of all-glass impingers. *AIHA*, **58** 480-88.
Lindahl, R., Levin, J-O. and Martensson, M. (1996) Validation of a diffusive sampler for the determination of acetaldehyde in air. *Analyst*, **121** 1177-81.
Ludwig, U., Grischek, T., Neitzel, P. and Nestler, W. (1997) Ultrafiltration: a technique for determining the molecular mass distribution of group parameters of organic pollutants in natural waters. *J. Chromatogr. A.*, **763** 315-21.
Maitre, A., Leplay, A., Perdix, A., Ohl, G., Boinay, P., Romanzini, S. and Aubrun, J.C. (1996) Comparison between solid sampler and impinger for evaluation of occupational exposure to 1,6-hexamethylene diisocyanate polyisocyanate during spray-painting. *AIHA*, **57** 153-60.
Malpei, F., Rozzi, A., Colli, S. and Uberti, M. (1997) Size distribution of TOC in mixed municipal-textile effluents after biological and advanced treatment. *J. Membr. Sci.*, **131** 71-83.
Manahan, S.E. (1991) *Environmental Chemistry*, 5th edn, Lewis Publishers, Michigan, p. 460.
Martinez, R.C., Gonzalo, E.R., Hermandez, F.E. and Hernandez, M.J. (1995) Membrane extraction-preconcentration cell coupled on-line to flow-injection and liquid chromatographic systems. Determination of triazines in oils. *Anal. Chim. Acta*, **304** 323-32.
Meares, P. (1986) The physical chemistry of transport and separation by membranes. In *Membrane Separation Processes* (ed. P. Meares), Elsevier Scientific Publishing Co., New York, USA, pp. 2-3.
Minuth, T., Thömmes, J. and Kula, M-R. (1996) A closed concept for purification of the membrane-bound cholesterol oxidase from *Nocardia rhodochrous* by surfactant-based cloud-point extraction, organic-solvent extraction and anion-exchange chromatography. *Biotechnol. Appl. Biochem.*, **23** 107-16.
Mitra, S., Zhu, N., Zhang, X. and Kebbekus, B. (1996) Continuous monitoring of volatile organic compounds in air emissions using an on-line membrane extraction-microtrap gas chromatographic system. *J. Chromatogr. A.*, **736** 165-73.

Munro, T.A. and Smith, B.D. (1997) Facilitated transport of amino acids by fixed-site jumping. *Chem. Comm.*, **22** 2167-68.

Notø, H., Hlagard, K., Daae, H.L., Bentsen, R. and Eduard, W. (1996) Comparative study of an inhalable and a total dust sampler for personal sampling of dust and polycyclic aromatic hydrocarbons in the gas and particulate phase. *Analyst*, **121** 1191-96.

Nriagu, J.O. (1992) *Gaseous Pollutants: Characterization and Cycling*. John Wiley and Sons, New York, USA, pp. 146-48.

Oms-Molla, M.T. and Klockow, D. (1996) Generation of test atmospheres of 2,3-dinitrophenol and evaluation of sampling methods for its determination in air. *Intern. J. Environ. Anal. Chem.*, **62** 281-88.

Ozadali, F., Glatz, B.A. and Glatz, C.E. (1996) Fed-batch fermentation with and without on-line extraction for propionic and acetic acid production by *Propionibacterium acidipropionici*. *Appl. Microbiol. and Biotechnol.*, **44** 710-16.

Peters, G.D., Hayes, J.M. and Hieftje, G.M. (1974) *Chemical Separations and Measurements: Theory and Practice of Analytical Chemistry*, W.B. Saunders, London, UK, p. 489 and p. 506.

Pinder, A.R. (1964) *Physical Methods of Organic Chemistry*, English Universities Press, London, UK, pp. 10-11.

Possanzini, M., Febo, A. and Liberti, A. (1983) New design of a high-performance denuder for the sampling of atmospheric pollutants. *Atmos. Environ.*, **17** 2605-10.

Possanzini, M., Ciccioli, P., di Palo, V. and Draisci, R. (1987) Determination of low boiling aldehydes in air and exhaust gases by using annular denuders combined with HPLC techniques. *Chromatographia*, **23** 828-34.

Price, R.M. (1997) Internet site; http://sunset.backbone.olemiss.edu/~cmprice/lectures/extract.html.

Reeve, R.N. (1994) *Environmental Analysis*. ACOL, John Wiley and Sons, Chichester, UK. p. 150.

Risse, U., Flammenkamp, E. and Kettrup, A. (1996) Determination of aromatic hydrocarbons in air using diffusion denuders. *Fres. J. Anal. Chem.*, **356** 390-95.

Rudzinski, W.E., Sutcliffe, R., Dahlquist, B. and Key-Schwarz, R. (1997) Evaluation of tryptamine in an impinger system and on XAD-2 for the determination of hexamethylene-based isocyanates in spray-painting operations. *Analyst*, **122** 605-608.

Schwarzenbach, R.P., Gschwend, P.M. and Imboden, D.M. (1993) *Environmental Organic Chemistry*, John Wiley and Sons, Chichester, UK.

Skoog, D.A., West, D.M. and Holler, F.J. (1992) *Fundamentals of Analytical Chemistry*, 6th edn, Saunders College Publishing, USA, p. 784.

Spangler, M.D. and Sidhom, M.B. (1996) Quantitation of the organic solvent extractables (OSE) of petrolatum and analysis by capillary gas chromatography. *J. Pharm. Biomed. Anal.*, **15** 139-43.

Spanne, M., Tinnerberg, H., Dalene, M. and Skarping, G. (1996) Determination of complex mixtures of airborne isocyanates and amine. Part 1. Liquid chromatography with ultraviolet detection of monomeric and polymeric isocyanates as their dibutylamine derivatives. *Analyst*, **121** 1095-99.

Streicher, R.P., Kennedy, E.R. and Lorbeau, C.D. (1994) Strategies for simultaneous collection of vapours and aerosols with emphasis on isocyanate sampling. *Analyst*, **119** 89-97.

Takeshita, R., Akinoto, Y. and Nito, S. (1995) Effective sampling system for polychlorinated dibenzo-*p*-dioxins and polychlorinated dibenzofurans in flue gas from municipal solid incinerators. *Environ. Sci. Technol.*, **29** 1186-94.

Thorpe, J.F. and Whitely, M.A. (1940) *Thorpe's Dictionary of Applied Chemistry*, 4th edn, Vol. iv., Longmans, Lowe and Brydone Ltd, London, UK, p. 562.

Ulicky, L. and Kemp, T.J. (1992) *Comprehensive Dictionary of Physical Chemistry*, Ellis Horwood, Chichester, UK, p. 148.

Warn, J.R.W. (1969) *Concise Chemical Thermodynamics*, Van Nostrand Reinhold Co., London, UK.

Wieczorek, P., Jonsson, J.A. and Mathiasson, L. (1997) Concentration of amino acids using supported liquid membrane with di-2-ethylhexyl phosphoric acid as a carrier. *Anal. Chim. Acta.*, **346** 191-206.

Wischmann, H., Steinhart, H., Hupe, K., Montresori, G. and Stegmann, R. (1996) Degradation of selected PAHs in soil/compost and identification of intermediates. *Intern. J. Environ. Anal. Chem.*, **64** 247-55.

Wu, W.S., Szklar, R.S. and Smith, R. (1997) Application of tryptamine as a derivatising agent for the determination of airborne isocyanates. Part 7. Selection of impinger solvents and the evaluation against dimethyl sulfoxide used in US NIOSH regulatory method 5522. *Analyst*, **122** 321-23.

Yurtov, E.V. and Koroleva, M.Y. (1996) Emulsions for liquid membrane extraction: properties and peculiarities. in *Chemical Separations with Liquid Membranes*. American Chemical Society, Washington, DC, USA, pp. 89-102.

Zander, A.K. and Pingert, P. (1997) Membrane-based extraction for detection of tastes and odours in water. *Water Res.*, **31** 301-309.

3 Solid phase extraction (SPE) in organic analysis
Alan J. Handley and R.D. McDowall

3.1 Introduction

Solid Phase Extraction (SPE) as a technique for sample isolation and concentration came to prominence in the early 1970s with the development of small, disposable cartridge systems containing solid absorbents, which could greatly speed-up the 'extraction' process prior to analysis. The traditional approach for solvent exchange was liquid/liquid extraction (LLE), in which the sample and an immiscible liquid were manually shaken and allowed to separate in a funnel. Since the two liquids must be immiscible, the range of properties shown by the extracting solvent is limited by the sample solvent, and, frequently, by the formation of intractable emulsions during the process, which can make further separation difficult. In addition, LLE can be extremely time-consuming and labour-intensive, and often requires large amounts of both extraction solvent and sample.

As a technique, SPE offers fewer disadvantages as the compounds are extracted from the sample onto a sorbent material contained in a column. Interferents can be selectively washed from the column and the desired analyte recovered by further solvent elution. Alternatively, the column can be operated to retain the interferents and allow the analyte of interest to pass through unretained.

Once a suitable combination of sorbent and solvent has been identified, the process can easily be automated to provide fractions suitable for further analysis by either High-Pressure Liquid Chromatography (HPLC), Gas Chromatography (GC) or Mass Spectroscopy (MS).

Whilst SPE is capable of being used for gases, solids and liquids, its primary area of application is in the selective extraction and enrichment of liquid samples. It is used extensively in the environmental, pharmaceutical, clinical, food and beverage and forensic sciences areas.

3.2 The background and principles of solid phase extraction

Solid phase extraction evolved from the early concepts of classical liquid solid extraction (LSE). In LSE, the extraction protocol consists of putting the sorbent into a vessel containing the analyte dissolved in a suitable solvent, and shaking for a controlled length of time. The two phases are then separated by filtration or decantation. If LSE has been successful, the analyte of interest will have been completely absorbed onto the solid phase (sorbent) to be subsequently desorbed using an appropriate solvent. By choosing suitable sorbent/solvent systems, selective elution, selective retention or unselective retention/elution can be effected (Figure 3.1).

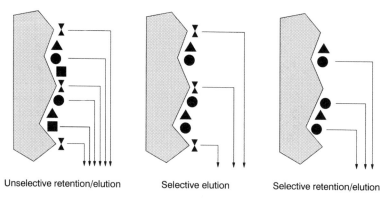

Figure 3.1 The liquid solid extraction (LSE) process.

This was later refined by packing the sorbent into glass columns to perform low efficiency liquid chromatography. Further developments saw the emergence of SPE, which is seen to separate the different compounds in a sample by utilizing the principles of modern liquid chromatography, both HPLC and Thin Layer Chromatography (TLC).

In SPE, the sample passes over the stationary phase, the analytes being separated according to the degree to which each component is partitioned or adsorbed by the stationary phase. The objective of an extraction scheme is to achieve 'digital chromatography' (McDowall et al., 1984); this is different from the normal aims of a chromatographer, who requires good peak shape and relatively short retention times. Extraction schemes based on chromatographic principles should either retain the analyte on the phase, allowing isolation and clean-up, or rapidly elute it in the smallest possible volume prior to analysis. In other words, the capacity factor, k', which in elution chromatographic analysis should be in the range 1–10, should be >1000 for retention and <0.001 for elution when using LSE to isolate molecules. Early developments centred around the use of adsorbents such as carbon, Celite® and alumina (Meala and Vanko, 1974; Hackett and Dusci, 1977; Anton and Sayre, 1962), but porous and bonded silicas were developed for the liquid chromatography market and rapidly became the sorbents of choice.

This explosion in HPLC and the associated rapid developments in silica technologies advanced the development of simpler, improved methodologies for SPE, resulting in the 'SPE cartridge'. The concept, similar to low-pressure liquid chromatography, consisted of packing milligram quantities of adsorbent between two fritted disks in a small polypropylene or polyethylene 'syringe type' cartridge (Figure 3.2), with the liquid phase being passed through under low pressure or vacuum.

Resin-based sorbents were also developed, ion-exchange (Dole et al., 1966) and non-ionic (Fujimoto and Wang, 1970), fuelled by requirements in the fields of pharmaceutical and biomedical science. Further refinements of the technology resulted in the development of 'extraction disks' for SPE (Figure 3.2). Here, the sorbent is supported on a membrane or disk. The particle chemistry is equivalent to that in conventional SPE cartridges; however, smaller particle size packings are possible. Since the sorbents used are similar to those in HPLC, all chemistries used in modern packing technologies can be applied to prepare liquid chromatography (LC) packing in forms suitable for SPE. Hence, the basic separation/retention mechanisms for LC are the basis of those for SPE (Moors et al., 1994).

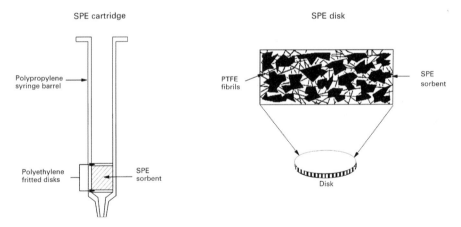

Figure 3.2 Solid phase extraction (SPE) technology. Abbreviation: PTFE, polytetrafluoroethylene.

The three main modes of separation are: normal phase, using polar adsorbents; reverse phase, with nonpolar bonded sorbents; and ion-exchange, using charged bonded sorbents. The mechanisms for separation depend primarily on the sorbent surface and, therefore, to understand the retention mechanisms involved in an extraction it is necessary to discuss the nature of the bonded surface (Burke et al., 1997). If we look at silica, the structure of the silica binds a permanent layer of water molecules, which is only removed by heating (Unger, 1979); protruding from the water layer are the bonded silica chains and interdispersed between them are molecules of water and the conditioning solvent (see Section 3.6). This configuration is controlled by solvation; the chains are more extended the higher the organic content (Unger and Lork, 1988). This swelling and shrinking behaviour will occur during the application and elution stages of the extraction. Also present on this surface, are residual silanols, which can play a significant role in the extraction scheme.

The analyte can be bound to the solid phase by a number of different mechanisms which are the same as LC, i.e. hydrogen-bonding, dipole-dipole interactions, hydrophobic dispersion forces and electrostatic (ionic) interactions (Kaliszan, 1987). Any or all of these forces will be involved during an extraction; it is the mastery of these forces that will determine the specificity of the method developed.

The energies in the many forces involved in bonding vary considerably. Hydrophobic bonding energies (dipole-dipole, dipole-induced dipole and dispersive interactions) range 1–10 kcal/mole. Hydrogen-bonding between suitable polar groups involves 5–10 kcal/mole but there are many opportunities for this type of interaction on the silica surface. Ionic or electrostatic interactions between oppositely charged species involve energies of 50–200 kcal/mole. It has been suggested that a more selective extraction could possibly be the result of utilising higher energy interactions than would be less likely to be formed. To a certain extent, this idea is borne out of the general observation that the selectivity of an extraction rises from C18 to C2 to the more polar phases, such as diol, cyanopropyl (CN) and amine.

It should also be noted that, even under the most rigorous end-capping procedures, it is not possible to end-cap all the silanol groups remaining after synthesis of the bonded phase; therefore, it is probable that silanol groups may be involved in isolation of an analyte. The pKa of a silanol group is not easily determined as it is influenced by the surrounding environment; however, at pH values of 2 the silanol is uncharged (Unger, 1979). Above this value it becomes increasingly dissociated and able to influence an extraction by virtue of its negative charge; electrostatic interactions involve higher energy than hydrophobic interactions. Therefore, if a mixed retention mechanism is present, the analyst must know the steps that can be taken to reduce or enhance the influence of residual silanols depending on the extraction mechanism desired. By careful control of these processes, the interactions can be tuned to give the surface a specific activity for a particular type or range of compounds, bringing the separating power and speed of liquid chromatography to sample extraction/preparation.

3.3 Instrumentation for solid phase extraction

A typical SPE system consists of the separating cartridge/column or disk and a means of facilitating flow of both solvent and sample through the sorbent. This is usually carried out by one of three methods (Figure 3.3): (i) pressure applied to the column inlet, which in its simplest form could be gravity or via a syringe; (ii) vacuum applied to the column outlet; or (iii) centrifugation.

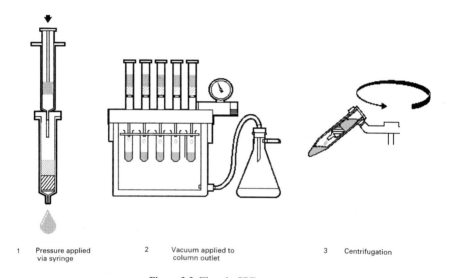

1 Pressure applied via syringe   2 Vacuum applied to column outlet   3 Centrifugation

Figure 3.3 Flow in SPE systems.

The above methods can be performed manually, however, with the widespread acceptance of the technique in the areas of both pharmaceutical and environmental science, systems based on both vacuum and syringe pumps have been developed for simultaneous (usually 10–20) multiple extractions (McDowall et al., 1986).

For high sample numbers or large volume extractions (water and liquid effluents), systems have been further automated to cover the whole SPE process.

These instruments are often based on robotic liquid handling systems (Figure 3.4). They can provide rapid SPE (Figure 3.5) and, in addition, can easily be interfaced to techniques such as HPLC for cost-effective sample preparation/extraction/analysis and improved precision and accuracy over manual methods.

In such systems, the SPE columns are capped to maintain the positive pressure required to flow the sample and solvents through the system and the wide-mouth cartridge geometry used (see Section 3.4.1).

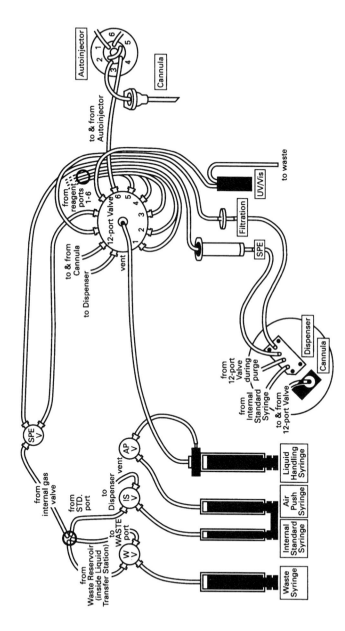

Figure 3.4 Zymark Benchmate II Workstation. Abbreviations: SPE, solid phase extraction; WS, waste syringe; IS, internal standard; AP, air push. (Courtesy of Zymark Corp., Zymark Center, Hopkinton, MA, USA).

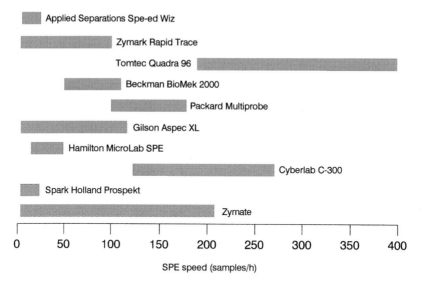

Figure 3.5 Throughput of automated SPE systems. (Smith and Lloyd, 1998).

3.4 The solid phase extraction column/cartridge and disk

3.4.1 The SPE column/cartridge

As stated previously, the column or cartridge is the key to the system. They come in many sizes and forms (Figure 3.6).

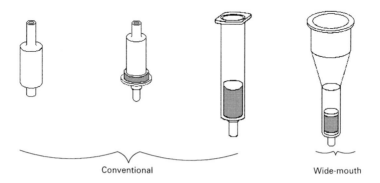

Figure 3.6 Types of solid phase extraction (SPE) cartridge.

The column is usually made from polypropylene with polyethylene frits to retain the sorbent. However, columns/cartridges made from either glass or virgin polytetrafluoroethane (PTFE) containing stainless steel and titanium frits have recently been developed for high sensitivity applications, such as GC analysis involving the use of electron capture and MS detection, where residual components extracted from the materials of construction can present problems (Junk *et al.*, 1988).

Cartridge/column volumes vary between 1 and 70 ml, but larger reservoir cartridges/adapters are available. Typical sorbent masses are between 50 mg and 10 g. The weight of the sorbent phase in the cartridge will determine the sample capacity; typically, the capacity in mg is equal to 5% of the sorbent mass, with the bed volume equal to 120 μl/100 mg of sorbent. The most frequently used cartridge sizes are 100 mg, 500 mg and 1 g (Majors, 1997). The packing materials are normally irregular in shape, between 40 and 50 μm mean particle diameter, and with a pore size of 60 A.

3.4.2 *The SPE disk*

The extraction disks closely resemble membrane filters; they are flat, usually 0.5–1 mm in thickness, with diameters ranging 4–96 mm. They can be membrane-like, in which the sorbent is either embedded or impregnated onto a flexible PTFE network, with as much as 90% by weight of silica being contained in the support (Figure 3.2). Alternatively, they can be of the rigid type, which consists of a fibreglass disk that contains the sorbent material.

The disks are supplied either loose, where users can install them into filter holders, or in preloaded disposable holders with Luer fittings. As with SPE cartridges, extraction disks can be used with all current forced-flow techniques (vacuum manifold, positive pressure and centrifuge), and are fully compatible with automated and robotic sample preparation workstations.

The SPE disks, because of their higher cross-sectional areas and shorter bed depths, are capable of being operated at much higher flow rates than SPE cartridges (1 l of water can pass through a 45 mm disk in approximately 10 min compared to 2 h when using a comparable SPE cartridge) (Majors, 1995). One further advantage of the SPE disk is that the sorbent particles are held tightly within the support, so that they are not susceptible to solvent/sample 'channelling', which can occur in packed-bed cartridges.

Clearly, there are both physical and chemical differences between disk and cartridge technologies (Blevins and Schulthesis, 1994); however, both offer the benefit of selective, automated extraction.

3.5 Solid phase extraction sorbents

SPE uses most of the phases and separation mechanisms available for HPLC; it should, however, be noted that there are less phases available in the disk format than the cartridge or column. The choice of cartridge or column for SPE depends on the nature of both the analyte and the matrix. Table 3.1 provides a general guideline for column selection.

In the main, most extractions carried out in SPE involve the use of bonded silicas as they offer both the necessary strength and rigidity and a wide range of solvent compatibility.

Three main separation mechanisms predominate when using such packings:

(i) Reverse phase or nonpolar extraction; cartridges used include octadecyl, octyl, cyclohexyl, butyl, phenyl and cyanopropyl. In such extractions, the sorbent used is less polar than the solvent or sample solution; the nonpolar sample constituents are retained by the bonded phase and eluted with nonpolar solvents. Moderately polar solvents may also be retained and eluted with more polar solvents.

(ii) Normal phase or polar extraction; cartridges used include silica, cyanopropyl, diol and aminopropyl. Here, polar sample constituents are preferentially retained and elution is achieved with relatively polar solvents. Normal phase extractions are best employed in situations involving a polar or moderately polar analyte in a nonpolar matrix. Other packings offering similar characteristics include alumina and Florisil (powdered magnesium silicon oxide).

(iii) Ion-exchange; cartridges can contain either strong or weak cationic and anionic exchangers, they can be either silica- or resin-based. In ion-exchange separations, cationic or anionic analytes in the sample are retained by an oppositely charged group on the exchanger. Elution generally occurs by changing the pH or increasing the ionic strength of the eluent.

Dual phases are also available, which come in three types: mixed mode, layered and stacked. Mixed mode phases are two chemically bonded phases of differing type (e.g. C8 and a cation exchange) which are mixed together in the same cartridge (Hajou *et al.*, 1996). Layered phases also combine two phases, but one is packed on top of the other. Stacked phases use two different cartridges in series (Mann and Burke, 1997).

All three types are useful in extending the range of a single phase and providing enhanced selectivity. Other more specific phases are available

Table 3.1 Column selection in solid phase extraction (SPE)

	Cartridge type	Typical analytes	Matrix	Typical solvents
Reverse phase or nonpolar extraction	C_{18} Octadecyl C_8 Octyl C_2 Ethyl CH Cyclohexyl PH Phenyl CN Cyanopropyl (end-capped)	Nonpolar functional groups, such as alkyl and aromatics	Polar solutions Water Buffers Biological fluids	Methanol Acetonitrile Ethyl acetate Chloroform Acidic methanol Hexane Methylene chloride
Normal phase or polar extraction	CN Cyanopropyl Diol Sl Silica Aminopropyl	Hydroxyls, amines hetero-atoms (S, O, N)	Nonpolar Solvents Hexane Oils Chloroform Lipids	Methanol Isopropanol Acetone Methylene chloride Ethyl acetate
Ion-exchange Cation-exchange extraction	SCX Benzenesulphonic acid PRS Propylsulphonic acid CBA Carboxylic acid	Bases, such as amines, pyrimidines	Water Low ionic strength Acidic buffers	Alkaline buffers High ionic strength Buffer Acetate, citrate and phosphate
Anion-exchange extraction	SAX Benzenesulphonic acid PSA Primary/secondary amine NH_2 Aminopropyl DEA Diethylaminopropyl	Acids, such as carboxylic acids, sulphonic acids, phosphates	Water Low ionic strength Alkaline buffers Biological fluids	Acidic buffers High ionic strength Buffers Phosphate and acetate

for use in 'affinity' (Brandsteterova *et al.*, 1996) and 'size exclusion' modes; these are predominately resin-based. Most of the nonpolar phases, unlike the polar and ion-exchange sorbents, are end-capped. The degree of end-capping can be important, as the residual silanols can often play an important role in retention and elution of compounds. The porosity and particle size distribution of the packing material may also be factors influencing selectivity of a separation.

In the main, the C18 octadecyl as with HPLC is by far the most widely-used phase. One point that should be noted, however, is that the same packing material from different manufacturers can have an effect on the selectivity of an assay. Ruane and Wilson (1987) tested the ability of C18 cartridges from two manufacturers to extract four β-blocking drugs. With two of the drugs tested, there were large differences in recovery between the two makes; this may provide a valuable source of selectivity when developing assays. The selectivity may depend, in part, on: the original silica particle; the extent and coverage of the bonded silica phase; the production methods; and the number of silanols left after end-capping.

Modern SPE advocates the use of disposable columns/cartridges. Some publications (Juergens, 1986) mention the reuse of solid-phase cartridges up to 10 times; this practice, often applied for economic reasons, must be carefully evaluated so that the results generated are not compromised. Reuse of bonded phase silica will inevitably affect its selectivity by coating of the bonded phase with endogenous constituents, making it similar to the precolumn in column-switching.

3.6 Strategies for solid phase extraction

3.6.1 *The SPE process*

The SPE process involves four key steps (Figure 3.7)

(i) *Column conditioning/solvation/equilibration*
Wetting of the column with an organic solvent, e.g. methanol, acetonitrile. This opens up the hydrocarbon chains and, thus, increases the surface area available for interaction with the analyte. In addition, it removes residues from the packing material that might interfere with the analysis. Failure to carry out this stage effectively will result in poor recoveries of analyte due to reduced retention on the column, and also interference peaks on the chromatogram which are unrelated to the original sample.

Washing of the sorbent bed with suitable solvent, e.g. HPLC grade water, buffer, etc. This will remove excess methanol or

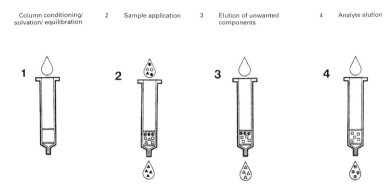

Figure 3.7 Key steps in the solid phase extraction (SPE) process.

similar solvents and prepare the surface for the sample. This conditioning step should be as similar as possible in polarity, ionic strength and pH value to the sample to be extracted. It should not be necessary to use a large volume of solvent; three or four times the bed volume of the cartridge is usually sufficient.

(ii) *Sample application*

Applying the sample, allowing it to flow through the sorbent bed and discarding the waste. Note that for samples containing small amounts of suspended solids or where the sample viscosity is high, it is often advantageous to use a larger diameter cartridge or column. Alternatively, as with the case of biological samples which are generally viscous, the samples can be diluted to speed the passage through the sorbent bed.

Blood samples can be assayed by SPE if the erythrocytes are lysed and the sample is centrifuged to remove the membranes. Solid samples that cannot be dissolved in the initial extract can be homogenized and filtered prior to extraction.

If a large volume of sample is used, the column may no longer be 'wetted' and a reduction in recovery will be observed. To overcome this problem, an organic solvent, e.g. methanol (1–3%) should be added to a sample of large volume prior to processing. This will help to maintain the equilibrium between the stationary and mobile phases. Flow rate can vary from cartridge to cartridge as the result of a number of factors, such as particle size distribution or the packing characteristics of individual columns. In some applications, it may be necessary to stop the flow through individual cartridges with the use of commercially available stopcocks to prevent the sorbent bed drying out. Flow rates through the sorbent bed should be controlled; based on personal experience, a maximum flow generated by creating a vacuum is recommended. If an analyte is strongly protein bound, then slower flow rates may be necessary to achieve a good recovery of the analyte.

(iii) *Elution of unwanted omponents*
Washing the column with water or a suitable solvent to selectively remove endogenous compounds from the sample matrix which might interfere with the subsequent analysis.

(iv) *Analyte elution*
Eluting the sample with a suitable solvent and collecting the eluent for immediate analysis or further work-up. The volume of the eluting solvent should be as small as possible to avoid dilution of the extract and lower limits of sensitivity. If the retention/elution mechanism is known, it should be possible to dilute the analyte in a very small volume, depending on the bed volume of the cartridge. Two dilutions of 100 µl are generally more efficient than a single application of 200 µl. A typical minimum elution volume is 250 µl/100 mg of sorbent. Flow control in the final step can be important, as excessive flow rates can result in reduced recoveries of analyte.

3.6.2 Method development in SPE

Method development must take into account: the nature of the analytes to be extracted; the sample matrix; the degree of analyte concentration required; the degree of purification required; the nature of the major contaminants in the sample; and the analytical detection method.

Certain steps can be taken to simplify the process.

(i) Searching the literature for chromatographic assay methods published for similar structures. If the analyte is a novel compound, it may be possible to deduce chromatographic conditions from similarities in structure to existing chemical entities.

(ii) Considering the physicochemical properties of the analyte, e.g. solubility and ionisation; this knowledge should give the conditions for a potential, retention mechanism, such as ion-exchange or hydrophobic on a particular type of phase.

(iii) Considering the physicochemical properties of the sample matrix: polar or nonpolar. This, combined with the analyte conditions, should help in further developing the extraction scheme.

(iv) Considering the properties of the HPLC mobile phase. This will give information about the elution conditions for the compound from the solid phase cartridge.

(v) Evaluating the retention of the analyte on various bonded phases if the HPLC separation is reversed-phase. It is important to remember that the pH can have a profound effect on retention of a compound (see (ii) above). In general, a stationary phase of similar polarity to the compound of interest is used, with the sample dissolved (and applied to the column) in a solvent of opposite polarity. The analyte is eluted by again switching solvent polarity. Thus, a relatively lipophilic organic molecule would be applied to a C18, C8 or C2 cartridge in a polar solvent, and eluted with a relatively nonpolar solvent.

(vi) In the pharmaceutical field, there is a need to compare the recoveries from aqueous and biological matrices. Differences in the behaviour of some drugs in these two matrices have been seen, for example oxmetidine has a recovery of > 90% from water but this falls to 50% in plasma due to competition from endogenous compounds. In contrast, failure to elute all the analyte from the phase after extraction from water (< 30% recovery) compared with > 95% recovery from plasma has been observed (Doyle et al., 1987). The masking of active silanols by components of plasma may be a possible explanation for this phenomenon.

(vii) Elution of the analyte from the cartridge must occur efficiently, so that the solution to be assayed is not too dilute.

(viii) The sorbent size can be optimised if required, although one group of workers prefer 100 mg size on the grounds of easier handling, smaller elution volumes and economy.

To evaluate any sample preparation scheme, a radiolabelled form of the compound can be used; the amount of radioactivity in the various fractions will quickly identify any problem areas of the potential extraction scheme. Alternatively, if radioactively-labelled analyte is not available, a solution should be passed down the columns and the eluent examined for presence of analyte. Those phases that retain the compound are then successively eluted with 1 ml portions of methanol, 0.1 M ammonia in methanol and 0.1 M acetic acid in methanol until elution occurs. Guidelines are available from the main vendors of these cartridges to assist in developing methods. The guidelines also include troubleshooting sections to help the analyst with any problems that may be encountered. Care must be taken to think the extraction through carefully, as the retention mechanism can often include secondary interactions which could affect the outcome of an extraction.

After the use of a solid phase extraction clean-up step, there may still be compounds from the sample matrix present in the eluate, which would interfere with the final measurement. One solution may be to vary the nature of the washing solvent to elute the unwanted compounds selectively. If this approach is unsuccessful, it may be necessary to use a second extraction column to provide a more efficient clean-up technique known as Chromatographic Mode Sequencing (CMS). An example of CMS is the analysis of tricyclic antidepressants in human plasma (Harkey and Stolowitz, 1984). The drug is retained on a C2 cartridge and interfering material is washed to waste with dilute sodium hydroxide. Chloroform is used to elute the drug from the C2 cartridge onto a CN cartridge, effecting a further purifications. The analytes are then eluted with HPLC mobile-phase and aliquots of this are taken for analysis. Alternatively, LLE can be combined with SPE.

3.7 Applications of solid phase extraction

As stated previously, there is a considerable data base of applications available from the SPE cartridge/column and disk suppliers (J. T. Baker Inc., 1995; Varian Associates, 1995; Waters Corp., 1995; 3M, 1995) and also in reference books (Thurman *et al.*, 1998; Simpson, 1997). Table 3.2 lists some of the many applications of SPE.

SPE has been used extensively in the pharmaceutical and biomedical fields for the extraction of a variety of matrices: antibiotics in serum (Cheng *et al.*, 1997); antioxidants in ointments (Nguyen, 1986); pharmaceutical creams (Bonazzi *et al.*, 1995); drugs in biological fluids (Cheng *et al.*, 1997), plasma (Brandstetrova *et al.*, 1995) and tissue (Brandstetrova *et al.*, 1996a); and in the bioanalysis of peptides (Causon and McDowall, 1992).

In the forensic area, it is used for screening for drugs of abuse and narcotics both in plasma (Schmid, 1989) and urine (Wolff, 1990). The area of environmental monitoring has seen a rapid increase in validated methods using SPE, with the Environmental Protection Agency (EPA) accepting SPE methodology as an alternative to LLE for a large number of analytes and matrices. These include the following analytes in drinking water: benzidines (EPA 553); carbonyl compounds (EPA 554); chlorinated pesticides (EPA 508.1); chlorinated acids (EPA 515.2); polycyclic aromatic hydrocarbons (EPA 550.1); diquat and paraquat (EPA 549.1); phthalates (EPA 506); tetrachlorodibenzo-*p*-dioxin (EPA 513); and extractable organics (EPA 525.1). In addition, for wastewaters, methods for organochlorine pesticides and polychlorinated biphenyls (EPA 3600), organophosphorus pesticides (EPA 1657) and phenoxy-acid herbicides (EPA 1658) have been accepted. Other analytes studied in water using SPE include sulphonated azo dyes (Schonsee *et al.*, 1997), aromatic amines and phenols (Piangerelli *et al.*, 1997), fluorinated aromatic

Table 3.2 Applications of solid phase extraction (SPE)

	Cartridge type	Typical application
Reverse phase or nonpolar extraction	Octadecyl (C_{18})	Drugs of abuse, acetaminophen, amines, analgesics, anthraquinones, antiarrhythmics, anticonvulsants, antiepileptics, antibiotics, antidiabetics, anti-inflammatory, aromatics, aromatic amines, barbiturates, benzodiazepines, biocides, caffeine, cannabis, carbohydrates, carboxylic acid, carotenoids, cholesterol esters, dye intermediates, essential oils, ethchlorvynol, ethosuximide, fatty acids, food preservatives, fungicides, herbicides, hydrocarbons, hypnotics, insecticides, lidocaine, lipids, oil soluble vitamins, phenols, phthalate esters, priority pollutants, pesticides (PNAs PAHs PCBs), sedatives, steroids, sulphonamides, surfactants, tetracyclines, theophylline, tricyclic antidepressants, triglycerides, valproic acid
	Octyl (C_8)	Priority pollutants (pesticides, PNAs PAHs PCBs), fungicides, herbicides, peptides, oligonucleotides
	Phenyl (C_6H_5)	Hydrophobic compounds
	Cyano (CN)	Amine benzyl alcohol, antidepressants, antifungals, beta-blockers, dye intermediates, hydrocarbons, oil soluble vitamins; pesticides (PAHs PCBs), phenols, sugar alcohols
Normal phase or polar extraction	Silica gel	Aflatoxins, alcohols, aldehydes, alkaloids, amines, amino acids amphetamines, antibiotics, antioxidants, aromatics, barbituates, carbohydrates, dye intermediates, flavinoids, herbicides, heterocyclic compounds, hydrocarbons, hydrocortisone, indoles, insecticides, ketones, lipids, nitrocompounds, oligopheylines, organic acids, peroxides, pesticides, phenols, plant pigments, plasticizers, polypeptides, porphyrins, steroids, terpenes, unsaturated compounds, fat soluble vitamins
	Diol (COHCOH)	Antibiotics, proteins, peptides, aqueous surfactants
	Amino (NH_2)	Carbohydrates, chlorophenols, fatty acids, food preservatives, lipids, metals, nucleotides, peptides, petroleum and lub oils, antidiabetics saccharides, steroids, sugars, vitamins
Ion-exchange		
Cation-exchange extraction	Aromatic sulphonic acid ($C_6H_5SO_3H$)	Amino acids, catecholamines, hormones, nucleic acid bases, nucleosides, purines, pyramidines, water soluble vitamins, chelating agents, dyes, herbicides
	Diamino	Amino acids
Anion-exchange extraction	Quaternary amine	Antibiotics, cyclic nucleotides, nucleotides, nucleic acids, bile acids, carboxylic acids

Abbreviations: PAHs, polycyclic aromatic hydrocarbons; PCBs, polychlorinated biphenyls, PNAs, polynuclear aromatic hydrocarbons.

carboxylic acids (Galdiga and Greibrokk, 1998) and microcontaminants (Rivasseau and Caude, 1995).

SPE has also found uses in the food and flavours technologies for extraction of components from wine (Kakalikova *et al.*, 1996), cereals (Biancardi and Riberzni, 1996), beer (Baltistutta *et al.*, 1994) and processed food (Hartmann *et al.*, 1996). Other less specific areas include dental materials (Shintani, 1992), fuels (Gehringer, 1996) and lubricants (Flake, 1992).

From the very volume of areas of application in the literature it can be seen that SPE has proved to be a valuable extraction technology for many areas of organic analysis.

3.8 New technologies in solid phase extraction

The power of SPE for simple and automated sample clean-up, concentration and matrix removal has resulted in the procedure becoming an integral part of analysis. More than 50 companies are now producing SPE-related products and, as such, the technology continues to evolve. Developments can be seen in a number of areas.

3.8.1 SPE phases

New sorbents continue to be produced for the column, cartridge or disk. New spherical polymeric phases have now been introduced which are less prone to fines, extractables, and phase shrinkage/swelling, and have greater water 'wettability' (Bouvier *et al.*, 1997). Designer phases are being developed, with the bonding technology being 'fitted' to the application (drugs of abuse, canine urine extraction), and targeted extractions/separations will be further exploited with the development of newer methods of preparing selective sorbents: restricted access (Oosterkamp, 1996) and molecular imprinting/recognition (Kriz *et al.*, 1997) media.

SPE product quality continues to improve as newer methods of characterising the sorbent materials are developed (Burke *et al.*, 1997), bringing closer the ultimate goals of high purity/reproducible phases.

The trend to use less solvent for extraction and smaller sample sizes is requiring the use of more sensitive detection systems for analysis (MS/MS). This places more emphasis on the construction materials of both the cartridges and phases, and the need to minimise/eliminate the possibility of extractables from such materials. Some of the newer packing materials now being used in HPLC—graphitised carbon and zirconium—may prove useful. SPE disk technology is continuing to develop, more sorbent types are becoming available, and the need for faster/high throughput extractions is resulting in smaller geometries, which can operate with much smaller volumes of elution solvent (see Section 3.8.2).

3.8.2 SPE formats

As observed above, the trend is to miniaturisation. In the pharmaceuticals field, advances in combinatorial chemistry and the associated requirements for high throughput screening of candidate compounds for development have resulted in the need for faster *in vivo* and *in vitro* extraction/screening methods. To serve this need, two developments have emerged, both designed to be accommodated in automated systems and linked to HPLC/MS/MS to provide fast selective and sensitive analysis.

The 96-well SPE extraction plate is based on either disk (Blevins and Hall, 1998) or packed-bed technologies (Allanson *et al.*, 1996; Kaye *et al.*, 1996) and will fit into any automated xyz liquid handling systems. The plate contains 96 individual SPE columns and is designed to sit on top of a similarly configured well collection system. With the growing success of such systems, 384-well formats and upwards are being considered.

A second development based on disk technology is the 'SPE pipette', which is a conventional pipette tip that is fitted with an SPE disk. This device can be fitted into a standard pipette tray, and can be operated either manually or in automated systems. The advantage of such a system is that it is capable of bidirectional flow. Liquids can be either drawn or forced through the disk, allowing for the possibility of backflush in either the clean-up or extraction process.

3.8.3 Linkage of SPE to other analytical techniques

Many of the developments in SPE have involved its linkage to HPLC for subsequent sample analysis and identification. Other techniques are now being readily interfaced to SPE. Links to other extraction techniques have been explored—to microwave (see Chapter 5) and to supercritical fluid extraction (Garimella *et al.*, 1996) to further enhance the separation process.

Direct interfacing to high sensitivity detectors (MS/MS), with SPE replacing the HPLC column, is gaining acceptance, particularly for high throughput screening of pharmaceutical compounds (Angelico *et al.*, 1997; Bowers *et al.*, 1997). Moreover, its ability to be used for sample concentration prior to chromatographic separation/determination has been further exploited in: SPE/GC/MS (Ollers *et al.*, 1997); SPE/GC

(Louter et al., 1997; Jahr et al., 1996); supercritical fluid chromatography (Bernal et al., 1997); and capillary electrophoresis (He and Lee, 1997).

Solid phase extraction continues to develop as an essential technique for sample extraction and concentration and, as can been seen from this chapter, it is very much suited to the requirements of the modern day chemist for fast, automated, miniaturised and environmentally friendly technology.

References

Allanson, J.P., Biddlecombe, R.A., Jones, A.E. and Pleasance, S. (1996) *Rapid Communications in Mass Spectrometry*, **10** 811-16.
Angelico, V.J., Mann, T.D., Burke, M.F. and Wysocki, V.H. (1997) *Abstracts of Papers of the American Chemical Society*, **214** Iss SEP, 21.
Anton, A.H. and Sayre, D.F. (1962) *J. Pharm. Exp. Ther.*, **138** 360.
Baltistutta, F., Buiatti, S., Zenarola, C. and Zironi, R. (1994) *HRC Journal of High Resolution Chromatogr.*, **17** 662-64.
Bernal, J.L., Nozal, M., Toribio, L., Borrull, F., Marce, R.M. and Pocurull, E. (1997) *J. Chromatogr.*, **778** 321-28.
Biancardi, A. and Riberzani, A. (1996) *Journal of Liquid Chromatography and Related Techniques*, **19** 2395-407.
Blevins, D.D. and Schulthesis, S.K. (1994) *LC-GC Magazine of Separation Science*, **12** 12.
Blevins, D.D. and Hall, D.O. (1998) *LC-GC Trends and Developments in Sample Preparation*, May, S22-S31.
Bonazzi, D., Andrisano, V., Gatti, R. and Cavrini, V. (1995) *J. Pharm. Biomed. Anal.*, **13** 1321-29.
Bouvier, E.S.P., Iraneta, P.C., Neue, U.D., Philips, D.J., Capperella, M. and Cheng, Y.F. (1997) Paper presented at the 124th American Chemical Society Meeting, Las Vegas, 12th Sept.
Bowers, G.D., Clegg, C.P., Hughes, S.C., Harker, A.J. and Lambert, S. (1997) *LC-GC Magazine of Separation Science*, **15** 48.
Brandsteterova, E., Kubalec, P., Rady, A. and Krcmery, V. (1995) *Pharmazie*, **50** 597-99.
Brandsteterova, E., Kubalec, P., Simko, P. and Machackova, L. (1996a) *Pharmazie*, **51** 984-86.
Brandsteterova, E., Kubalec, P., Krajnak, K. and Skacani, I. (1996b) *Neoplasma*, **43** 107-12.
Burke, M.F., Raisglid, M. and Piccoli, R.F. (1997) *Abstracts of Papers of the American Chemical Society*, **214** (Sep) 170.
Causon, R.C. and McDowall, R.D. (1992) *Journal of Controlled Release*, **21** 37-48.
Cheng, Y.F., Philips, D.J. and Neue, U.D. (1997) *Abstracts of Papers of the American Chemical Society*, **214** (Sep) 44.
Cheng, Y.F., Philips, D.J. and Neue, U.D. (1997) *Chromatographia*, **44** 187-90.
Dole, V.P., Lim, W.K. and Eglitis, I. (1966) *J. Am. Med. Assoc.*, **198** 349.
Doyle, E., Pearce, J.C., Picot, V.S. and Lee, R.M. (1987) *J. Chromatogr.*, **411** 325.
Flake, C.J. (1992) *LC-GC Magazine of Separation Science*, **10** 926.
Fujimoto, J.M. and Wang, R.I.H. (1970) *Toxicol. Appl. Pharmacol*, **16** 186
Garimella, U.I., Stearman, G.K. and Wells, M.J.M. (1996) *Abstracts of Papers of the American Chemical Society*, **211** (Mar) 194.
Galdiga, C.H. and Greibrokk, T. (1998) *J. Chromatogr. A*, **793** 297-306.
Gehringer, J.M. (1996) *Abstracts of the American Chemical Society*, **212** (Aug) 68.
Hackett, L.P. and Dusci, L.J. (1977) *J. Forensic Sci.*, **22** 376.

Hajou, Z., Thomas, J. and Souffi, A.M. (1996) *Journal of Liquid Chromatography and Related Technologies*, **19** 1937-45.
Harkey, M.R. and Stolowitz, M.L. (1984) *Advances in Analytical Technology* (ed. R.D. Baselt), Vol. 1, Biomedical Publications, Foster City, p. 255.
Harris, P.A. (1985) Proceedings of the 2nd International Symposium 'Sample Preparation and Isolation Using Bonded Silicas', Philadelphia, January, 14-15.
Harrison, A.C. and Walker, D.K. (1998) *J. Pharm. Biomed. Anal.*, **16** 777-83.
Hartmann, M., Ammon, J. and Berg, H. (1996) *Deutsche Lebensmittel-Rundschau*, **92** 137-41.
He, Y. and Lee, H.K. (1997) *Electrophoresis*, **18** 2036-41.
Jahr, D., Vreuls, J., Louter, A. and Loebel, W. (1996) *GIT Fashz. Lab.*, **40** (3) 178 and 180-83.
Juergens, U. (1986) *J. Chromatogr.*, **371** 307.
Junk, G.A., Avery, M.J. and Richard, J.J. (1988) *Anal. Chem.*, **60** 1347-50.
J.T. Baker Inc. (1995) *Bakerbond SPE Bibliography*.
Kaliszan, R. (1987) *Chemical Analysis in Quantitative Structure: Chromatographic Retention Relationships*, Vol. 93 (ed. J.D. Winefordner), Wiley, New York, p. 7.
Kakalikova, L., Matisova, E. and Lesko, J. (1996) *Zeitschrift für Lebensmittel-Untersuchung und Forschung*, **203** 56-60.
Kaye, B., Herron, W.J., Macrae, P.V., Robinson, S., Stopher, R.V. and Wild, W. (1996) *Anal. Chem.*, **58** 1658-60.
Kohler, J., Chase, D.B., Farlee, R., Vega, A.J. and Kirkland, J.J. (1986) *J. Chromatogr.*, **352** 275.
Kriz, D., Ramstrom, O. and Mosbach, K. (1997) *Anal. Chem.*, **11** 345A.
Louter, A.J.H., Jones, P.A., Jorritsma, D.J., Vereuls, J.J. and Brinkman, U.A.Y. (1997) *J. High Resolut. Chromatogr.*, **20** 363-68.
Majors, R.E. (1995) *LC-GC*, **8** 128-33.
Majors, R.E. (1997) *LC-GC*, **15** 310-24.
Mann, T.D. and Burke, M.F. (1997) *Abstracts of Papers of the American Chemical Society*, **214** (Sep) 22.
Meala, R. and Vanko, P. (1974) *Clin. Chem.*, **20** 184.
McDowall, R.D., Murkitt, G.S. and Walford, J.A. (1984) *J. Chromatogr.*, **317** 475.
McDowall, R.D., Pearce, J.C. and Murkitt, G.S. (1986) *J. Pharm. Biomed. Anal.*, **4** 3.
Moors, M., Massart, D.L. and McDowall, R.D. (1994) *Pure and Applied Chemistry*, **66** 277-304.
3M (ed.) (1995) *Bibliography of Empore Membrane Publications*.
Nguyen, T.T., Kringstad, R. and Rasmussen, K.E. (1986) *J. Chromatogr.*, **366** 445-50.
Oosterkamp, A.I., Irth, H., Heintz, L., Narko-Varga, G., Tjaden, U.R. and Vander Greef, J. (1996) *Anal. Chem.*, **68** 4101-106.
Ollers, S., Van Lieshout, M., Janssen, H.J. and Cramers, C.A. (1997) *LC-GC*, **15** 846.
Papadoyannis, I.N., Tsioni, G.K. and Samanidou, V.F. (1997) *Journal of Liquid Chromatography and Related Technologies*, **20** 3203-31.
Piangerelli, V., Nerini, F. and Cavalli, S. (1997) *Annali di Chimica*, **87** 571-82.
Rivasseau, C. and Caude, M. (1995) *Chromatographia*, **41** 462-70.
Runane, R.J. and Wilson, I.D. (1987) *J. Pharm. Biomed. Anal.*, **5** 723.
Schmid, R. and Kupferschmidt, R. (1989) *Clin. Chem.*, **35** (7) 1352.
Schonsee, I., Rui, J. and Barcelo, D. (1997) *Qium. Anal.*, **16** 243-49.
Shintani, H. (1992) *Journal of Liquid Chromatography*, **15** 1315-35.
Simpson, N. (ed.) (1997) *Solid Phase Extraction: Principles, Strategies and Applications*, Decker.
Smith, G.A. and Lloyd, T.L. (1998) Automated solid phase extraction and sample preparation. *LC-GC*, **May** 522-31.
Thurman, E.M. and Mills, M.S. (eds.) (1998) *Soild Phase Extraction: Principles and Practices*, John Wiley & Sons, New York.
Unger, K.K. (ed.) (1979) *Porous Silica*, Elsevier Amsterdam, p. 57.
Unger, K.K. and Lork, K.D. (1988) *Europ. Chromatogr. News*, **2** 14.

Varian Associates (ed.) (1995) *Applications Bibliography*.
Waters Corp. (1995) *Solid Phase Extraction Applications Guide and Bibliography* (eds. P.D. MacDonald and E.S.P. Bouvier).
Wolff, K., Sanderson, M.J. and Hay, A.W.N. (1990) *Ann. Clin. Biochem.*, **27** 482-88.

4 Solid phase microextraction (SPME) and membrane extraction with a sorbent interface (MESI) in organic analysis

Yuzhong Luo and Janusz Pawliszyn

4.1 Introduction

The ultimate goal of an analytical chemist is to perform an analysis at the place where a sample is located rather than moving the sample to a laboratory, as is common practice in many cases at present. This would eliminate errors and the time associated with sample transport and storage, and would therefore result in more accurate, precise and rapid analytical data collection. In addition to portability, two other important features of ideal field sample preparation technique are elimination of solvent use and integration with a sampling step. These requirements are met by two techniques based on fibre geometry and polymer technologies, which are discussed in this chapter. Solid Phase Microextraction (SPME) involves exposing a polymer-coated, fused silica fibre to a sample. The analytes partition into the stationary phase until an equilibrium has been reached, after which the fibre is removed from the solution and the analytes are thermally desorbed in the injector of an analytical instrument, such as a gas chromatograph. The fibre is contained in a syringe-like device to facilitate handling. SPME is capable of both spot and time-averaged sampling. Typically, for spot sampling, the fibre is exposed to a sample matrix until the partitioning equilibrium is reached between sample matrix and the coating material. In the time-averaged technique, the fibre remains in the needle during the exposure of the SPME device to the sample. The coating works as a trap for analytes which diffuse into the needle.

In Membrane Extraction with a Sorbent Interface (MESI), a polymeric hollow fibre membrane, which is in contact with the sample, is fitted directly into the carrier gasline of a gas chromatograph equipped with a sorbent trap. Analytes partition into the polymeric phase of the membrane and after diffusion through the membrane are carried by the gas to the sorbent trap. The concentrated analytes are periodically delivered onto the front of the column by a thermal pulse. MESI is a dynamic system, in which the rate of analyte intake is dependent both on the diffusion coefficients of the analytes in the membrane material and the distribution constant of the membrane/sample matrix. MESI, like SPME, can be used both for spot and time-averaged monitoring.

Both SPME and MESI techniques integrate sampling with sample preparation and introduction of the sample to the analytical instrument into a simple procedure. In SPME, mechanical movement of the fibre is necessary; however, sampling and sample introduction steps are separated in space allowing one instrument to analyse a large number of fibres. MESI, on the other hand, requires a dedicated instrument, permanently attached to one or several membrane/sorbent systems, but it eliminates the need for mechanical movement and, therefore, reduces the possibilities of failure.

4.2 Solid Phase Microextraction (SPME)

4.2.1 Introduction

This section introduces the basic concepts facilitating accurate and precise quantitation using SPME technology. The information presented below is a summary of the comprehensive discussion of the topic covered in a recently published book (Pawliszyn, 1997).

SPME was introduced as a solvent-free sample preparation technique in 1990. The basic principle of this approach is to use a small amount of the extracting phase (usually less than 1 μl) compared to the sample matrix. The sample volume can be very large, when the system investigated, e.g. room air or lake water, is sampled directly. The extracting phase can be either high molecular weight polymeric liquid, similar in nature to stationary phases in chromatography, or it can be a solid sorbent, typically of a high porosity to increase the surface area available for adsorption.

Up to the present time, the most practical geometric configuration of SPME has utilized a small, fused silica fibre, usually coated with a polymeric phase. The fibre is mounted for protection in a syringe-like device. The analytes are absorbed or adsorbed by the fibre phase (depending on the nature of the coating) until an equilibrium is reached in the system. The amount of an analyte extracted by the coating at equilibrium is determined by the magnitude of the partition coefficient (distribution ratio) of the analyte between the sample matrix and the coating material.

In SPME, analytes are typically not extracted quantitatively from the matrix. However, equilibrium methods are more selective because they take full advantage of the differences in extracting-phase/matrix distribution constants to separate target analytes from interferents. Exhaustive extraction can be achieved in SPME when the distribution constants are large enough. This can be accomplished for most compounds by the application of an internally cooled fibre (Zhang and Pawliszyn, 1995). In exhaustive extraction, selectivity is sacrificed to obtain quantitative transfer of target analytes into the extracting phase. One advantage of this approach is that, in principle, it does not require calibration, since all the analytes of interest are transferred to the extracting phase. On the other hand, the equilibrium approach usually requires calibration through the use of surrogates or standard addition to quantify the analytes and compensate for matrix-to-matrix variations and their effect on distribution constants.

Since equilibrium rather than exhaustive extraction occurs in the microextraction methods, SPME is ideal for field monitoring. It is unnecessary to measure the volume of the extracted sample and, therefore, the SPME device can be exposed directly to the system under investigation for quantitation of target analytes. In addition, extracted analytes are introduced to the analytical instrument by simply placing the fibre in the desorbtion unit (Figures 4.1B and C). This convenient, solvent-free process facilitates sharp injection bands and rapid separations (Gorecki and Pawliszyn, 1995). These features of SPME result in integration of the first steps in the analytical process: sampling; sample preparation; and introduction of the extracted mixture to the analytical instrument.

The equilibrium nature of the technique also facilitates speciation in natural systems, since the presence of a minute fibre, which removes only small amounts of target analytes, is not likely to disturb the system. Because of the small size, coated fibres can be used to extract analytes from very small samples. For example, SPME has been used to probe for substances emitted by a single flower bulb during its lifespan.

Figure 4.1A illustrates the commerical SPME device, manufactured by Supelco, Inc. (Bellefonte, PA, USA). The fibre, glued into a piece of stainless steel tubing, is mounted in a special holder. The holder is equipped with an adjustable depth gauge, which makes it possible to control repeatably how far the needle of the device is allowed to penetrate the sample container (if any) or the injector. This is important, as the fibre can easily be broken when it hits an obstacle. The movement of the plunger is limited by a small screw moving in the z-shaped slot of the device. For protection during storage or septum piercing, the fibre is withdrawn into the needle of the device, with the screw in the uppermost position. During extraction or desorption, the fibre is exposed by depressing the plunger, which can be locked in the lowered (middle) position by turning it clockwise (the position depicted in Figure 4.1A). The plunger is moved to its lowermost position only for replacement of the fibre assembly. Each type of fibre has a hub of a different colour. The

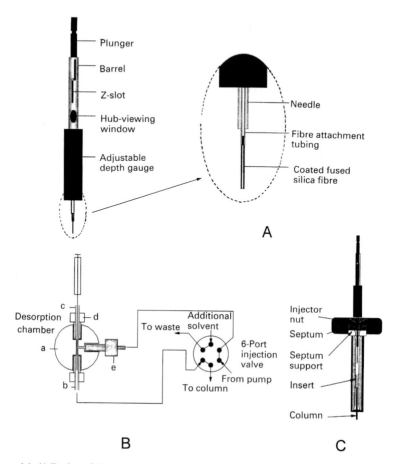

Figure 4.1 A) Design of the commercial solid phase microextraction (SPME) device. B) SPME/high-performance liquid chromatography (HPLC) interface: a, stainless steel (SS) 1/16 in tee; b, 1/16 in SS tubing; c, 1/16 in polyetheretherketone (PEEK) tubing (0.02 in ID); d, two-piece finger-tight PEEK union; e, PEEK tubing (0.005 in ID) with a one-piece PEEK union. C) SPME/gas chromatography (GC) interface.

hub-viewing window enables a quick check of the type of fibre mounted in the device.

If the sample is placed in a vial, the septum of the vial is first pierced with the needle (with the fibre in the retracted position), and the plunger is lowered, which exposes the fibre to the sample. The analytes are allowed to partition into the coating for a predetermined time, and the fibre is then retracted into the needle. The device is next transferred to the analytical instrument of choice. When gas chromatography (GC) is used for analyte separation and quantitation, the fibre is inserted into a hot injector, where thermal desorption of the trapped analytes takes place (Figure 4.1C). The process can be automated by using an appropriately modified syringe autosampler, commercially available from Varian (Varian Associates, Sunnyvale, CA, USA). For high-performance liquid chromatography (HPLC) applications, a simple interface mounted in place of the injection loop can be used to re-extract analytes into the desorption solvent (Figure 4.1B). HPLC interfaces are available from Supelco, Inc. (Bellefonte, PA, USA).

SPME sampling can be performed in three basic modes: direct extraction; headspace extraction; and extraction with membrane protection. Figure 4.2 illustrates the differences between these modes. In direct extraction mode (Figure 4.2A), the coated fibre is inserted into the sample and the analytes are transported directly from the sample matrix to the extracting phase. To facilitate rapid extraction, some level of agitation is required to transport the analytes from the bulk of the sample to the vicinity of the fibre. For gaseous samples, natural air flow (e.g. convection) is frequently sufficient to facilitate rapid equilibration. However, for aqueous matrices, more efficient agitation techniques, such as fast sample flow, rapid fibre or vial movement, stirring or sonication, are required to reduce the effect of the 'depletion zone' produced close to the fibre as a result of slow diffusional analyte transport through the stationary layer of liquid surrounding the fibre.

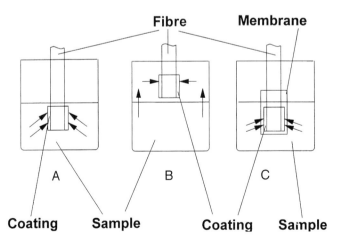

Figure 4.2 Modes of solid phase microextraction (SPME) operation. A) Direct extraction; B) headspace extraction; and C) membrane-protected SPME.

In the headspace mode (Figure 4.2B), the analytes are extracted from the gas phase equilibrated with the sample. The primary reason for this modification is to protect the fibre from adverse effects caused by nonvolatile, high molecular weight substances present in the sample matrix (e.g. humic acids or proteins). The headspace mode also allows matrix modifications, including pH adjustment, without affecting the fibre. In a system consisting of a liquid sample and its headspace, the amount of an analyte extracted by the fibre coating does not depend on the location of the fibre (in the liquid phase or in the gas phase), therefore, the sensitivity of headspace sampling is the same as the sensitivity of direct sampling, as long as the volumes of the two phases are the same in both sampling modes. Even when no headspace is used in direct extraction, a significant difference in sensitivity between direct and headspace sampling can occur only for very volatile analytes. However, the choice of sampling mode has a very significant impact on the extraction kinetics. When the fibre is in the headspace, the analytes are removed from the headspace first, followed by indirect extraction from the matrix. Therefore, volatile analytes are extracted faster than semivolatiles. Temperature has a significant effect on the kinetics of the process, since it determines the vapour pressure of the analytes. In general, the equilibration times for volatile compounds are shorter for headspace SPME than for direct extraction under similar conditions of agitation, because: (i) a substantial portion of the analytes is present in the headspace prior to the beginning of the extraction process; (ii) there is usually a large interface between the sample matrix and headspace; and (iii) the diffusion coefficients in the gas phase are typically higher by four orders of magnitude than in liquids. The concentration of semivolatile compounds in the gaseous phase at room temperature is small; headspace extraction rates for such compounds are substantially lower. They can be improved by using very efficient agitation or by increasing the extraction temperature.

In the third mode, SPME with membrane protection (Figure 4.2C), the fibre is separated from the sample by a selective membrane, which allows the analytes through while blocking the intereferents. The main purpose for the use of the membrane barrier is to protect the fibre against adverse effects caused by high molecular weight compounds when very dirty samples are analysed. While extraction from headspace serves the same purpose, membrane protection enables the analysis of less volatile compounds. The extraction process is substantially slower than direct extraction because the analytes need to diffuse through the membrane before they can reach the coating. Use of thin membranes and increased extraction temperature result in shorter extraction times.

4.2.2 Theoretical aspects of SPME optimization and calibration

4.2.2.1 Thermodynamics

SPME is a multiphase equilibration process. Frequently, the extraction system is complex, as in a sample consisting of an aqueous phase with suspended solid particles having various adsorption interactions with the analytes, plus a gaseous headspace. In some cases, specific factors have to be considered, such as loss of analytes by biodegradation or adsorption onto the walls of the sampling vessel. The discussion below will consider only three phases: the fibre coating; the gas phase or headspace; and a homogeneous matrix, such as pure water or air. During extraction, analytes migrate between all three phases until equilibrium is reached.

The mass of an analyte extracted by the polymeric coating is related to the overall equilibrium of the analyte in the three-phase system. Since the total mass of an analyte should remain constant during the extraction, then:

$$C_0 V_s = C_f^\infty V_f + C_h^\infty V_h + C_s^\infty V_s \tag{4.1}$$

where, C_0 is the initial concentration of the analyte in the matrix, C_f^∞, C_h^∞ and C_s^∞ are the equilibrium concentrations of the analyte in the coating, the headspace and the matrix, respectively, and V_f, V_h and V_s are the volumes of the coating, the headspace and the matrix, respectively. If the coating/gas distribution constant is defined as $K_{fh} = C_f^\infty / C_h^\infty$, and the gas/sample matrix distribution constant as $K_{hs} = C_h^\infty / C_s^\infty$, the mass of the analyte absorbed by the coating, $n = C_f^\infty V_f$, can be expressed as:

$$n = \frac{K_{fh} K_{hs} V_f C_0 V_s}{K_{fh} K_{hs} V_f + K_{hs} V_h + V_s} \tag{4.2}$$

Also:

$$K_{fs} = K_{fh} K_{hs} = K_{fg} K_{gs} \tag{4.3}$$

since, the fibre/headspace distribution constant, K_{fh}, can be appoximated by the fibre/gas distribution constant, K_{fg}, and the headspace/sample distribution constant, K_{hs}, by the gas/sample distribution constant, K_{gs}, if the effect of moisture in the gaseous headspace can be neglected. Thus, Equation 4.2 can be rewritten as:

$$n = \frac{K_{fs} V_f C_0 V_s}{K_{fs} V_f + K_{hs} V_h + V_s} \tag{4.4}$$

The equation states, as expected from the equilibrium conditions, that the amount of analyte extracted is independent of the location of the fibre in the system. It may be placed in the headspace or directly in the sample, as long as the volumes of the fibre coating, headspace and sample are kept constant. There are three terms in the denominator of Equation 4.4 which give measures of the analyte capacity of each of the three phases: fibre ($K_{fs} V_f$), head space ($K_{hs} V_h$), and the sample itself (V_s). If we assume that the vial containing the sample is completely filled (no headspace), the term $K_{hs}V_h$ in the denominator of Equation 4.4, which is related to the capacity ($C_h^\infty V_h$) of the headspace, can be eliminated, resulting in:

$$n = \frac{K_{fs} V_f C_0 V_s}{K_{fs} V_f + V_s} \qquad (4.5)$$

Equation 4.5 describes the mass absorbed by the polymeric coating after equilibrium has been reached in the system. In most determinations, K_{fs} is relatively small compared to the phase ratio of sample matrix to coating volume ($V_f \ll V_s$). In that situation, the capacity of the sample is much larger compared to the capacity of the fibre, resulting in a very simple relationship:

$$n = K_{fs} V_f C_0 \qquad (4.6)$$

The above equations emphasize the field sampling capability of the SPME technique. It is not necessary to sample a well-defined volume of the matrix, since the amount of analyte extracted is independent of V_s as long as $K_{fs} V_f \ll V_s$. The SPME device can be placed directly in contact with the system under investigation to allow quantitation.

Prediction of distribution constants. In many cases, the distribution constants present in Equations 4.2 to 4.6 which determine the sensitivity of SPME can be estimated from physicochemical data and chromatographic parameters. For example, distribution constants between a fibre coating and gaseous matrix (e.g. air) can be estimated using isothermal GC retention times on a column with a stationary phase identical to the fibre coating material. This is possible because the partitioning process in GC is analogous to the partitioning process in SPME, and there is a well-defined relationship between the distribution constant and the retention time. The nature of the gaseous phase does not affect the distribution constants unless the components of the gas, such as moisture, swell the polymer, thus changing its properties. A most useful method for determining coating-to-gas distribution constants uses the linear temperature-programmed retention index (LTPRI) system, which indexes the retention times of compounds relative to the retention times of n-alkanes. This sytem is applicable to retention times for temperature-programmed gas-liquid chromatography (GLC). The logarithm of the coating-to-air distribution constants of n-alkanes can be expressed as a linear function of their LTPRI values. For polydimethylsiloxane (PDMS), this relationship is log K_{fg} = 0.00415 * LTPRI−0.188 (Martos *et al.*, 1997). Thus, the LTPRI system permits interpolation of the K_{fg} values from the plot of log K_{fg} versus the retention index. The LTPRI values for many compounds are available in the literature, hence this method allows estimation of K_{fg} values without experimentation. If the LTPRI value for a compound is not available from published sources, it can be determined from a GC run. Note that the GC column used to determine LTPRI should be coated with the same material as the fibre coating.

Estimation of the coating/water distribution constant can be performed using Equation 4.5. The appropriate coating/gas distribution constant can be found by applying the techniques discussed above, and the gas/water distribution constant (Henry's constant) can be obtained from physicochemical tables or can be estimated by the structural unit contribution method (Martos *et al.*, 1997).

Some correlations can be used to anticipate trends in SPME coating/water distribution constants for analytes. For example, a number of investigators have reported the correlation between the octanol/water distribution constant K_{ow} and K_{fw}. This is expected, since K_{ow} is a very general measure of the affinity of compounds to the organic phase. It should be remembered, however, that the trends are valid only for compounds within homologous series, such as aliphatic hydrocarbons, aromatic hydrocarbons or phenols; they should not be used to make comparisons between different classes of compounds, because of different analyte activity coefficients in the polymer.

Effect of extraction parameters. Thermodynamic theory predicts the effects of modifying certain extraction conditions on partitioning and indicates parameters to control for reproducibility. The theory can be used to optimize the extraction conditions with a minimum number of experiments and to correct for variations in extraction conditions, without the need to repeat calibration tests under the new conditions. For example, SPME analysis of outdoor air may be performed at ambient temperatures that can vary significantly. The relationship that predicts the effect of temperature on the amount of analyte extracted allows calibration without the need for extensive experimentation (Schwarzenbach *et al.*, 1993). Extraction conditions that affect K_{fs} include: temperature, salting, pH, and the organic solvent content in water. A brief discussion about optimization of extraction parameters in the SPME method can be found in Section 4.2.2.2.

4.2.2.2 Kinetics

The kinetic theory is very useful in the optimization of extraction conditions by identifying 'bottlenecks' of SPME, and indicates strategies to increase extraction speed. In the discussion below, consideration will be limited to direct extraction (Figure 4.3).

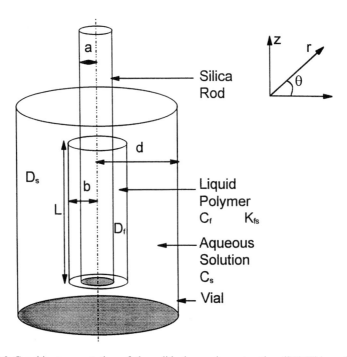

Figure 4.3 Graphic representation of the solid phase microextraction (SPME)/sample system configuration, with dimensions and parameters labelled as follows: a, inner radius of fibre coating; b, outer radius of fibre coating; L, length of fibre coating; d, inner radius of vial; C_f, concentration of analyte in the fibre coating; D_f, diffusion coefficient of analyte in the fibre coating; C_s, concentration of analyte in the sample; D_s, diffusion coefficient of analyte in the sample; K_{fs}, distribution coefficient of analyte between fibre coating and sample; $K_{fs} = C_f/C_s$.

Perfect agitation. Let us first consider the case where the liquid or gaseous sample is perfectly agitated. In other words, the sample phase moves very rapidly with respect to the fibre, so that all the analytes present in the sample have access to the fibre coating. In this case, the equilibration time, defined as the time required to extract 95% of the equilibrium amount (Figure 4.4) of an analyte from the sample, corresponds to:

$$t_e = t_{95\%} = \frac{2(b-a)^2}{D_f} \qquad (4.7)$$

Using this equation, one can estimate the shortest equilibration time possible for the practical system by substituting appropriate data for the diffusion coefficient of an analyte in the coating (D_f) and the fibre coating thickness ($b - a$). For example, the equilibration time for the extraction of benzene from a perfectly stirred aqueous solution with a 100 μm PDMS film is expected to be about 20 s. Equilibration times close to those predicted for perfectly agitated samples have been obtained experimentally for extraction of analytes from air samples (because of the high diffusion coefficients in gas) or when very high sonication power was used to facilitate mass transfer in aqueous samples. However, in practice, there is always a layer of unstirred water around the fibre. A higher stirring rate will result in a thinner water layer around the fibre.

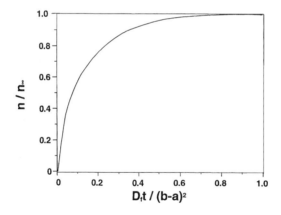

Figure 4.4 Mass absorbed versus time from perfectly agitated solution of infinite volume.

Practical agitation. Independent of the agitation level, fluid contacting a fibre's surface is always stationary, and as the distance from the fibre surface increases, the fluid movement gradually increases until it corresponds to bulk flow in the sample. To model mass transport, the gradation in fluid motion and convection of molecules in the space surrounding the fibre's surface can be simplified by a zone of a defined thickness in which no convection occurs, and perfect agitation in the bulk of the fluid everywhere else. This static layer zone is called the Prandtl boundary layer (Figure 4.5) (Martos and Pawliszyn, 1997); its thickness is determined by the agitation conditions and the viscosity of the fluid.

The equilibration time can be estimated, in practice, from the following equation:

$$t_e = t_{95\%} = 3 \frac{\delta K_{fs}(b-a)}{D_s} \qquad (4.8)$$

where, $(b-a)$ is the thickness of the fibre coating, D_s is the diffusion coefficient of the analyte in the sample fluid, and K_{fs} is the distribution constant of the analyte between the fibre and the sample. This equation can be used to predict equilibration times when the extraction rate is controlled by the diffusion in the boundary layer. In other words, the extraction time calculated by using Equation 4.8 must be longer than the corresponding time predicted by Equation 4.7.

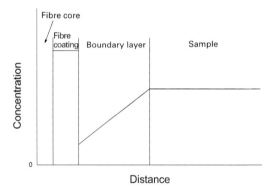

Figure 4.5 Boundary layer model configuration showing the different regions considered and the assumed concentration versus radius profile for the case when the boundary layer determines the extraction rate.

4.3 Membrane Extraction with Sorbent Interface (MESI)

4.3.1 Introduction

Membrane separation has been a rapidly developing field in recent years. Various membrane separation techniques have attracted increased attention and obtained wide application in trace organic monitoring and analysis. Numerous applications of membrane separation techniques are reported yearly. Membrane introduction mass spectrometry (MIMS) has been widely applied in air, water and biological analysis (Lotiaho et al., 1991; Bier and Cooks, 1987; Lauristen and Gylling, 1995; LapPack *et al.*, 1990; Virkki *et al.*, 1995). The supported liquid membrane (SLM) technique has been coupled with GC and LC for the analysis of ionizable and charged species (Jönsson and Mathiasson, 1992; Lindegård *et al.*, 1994; Thordarson *et al.*, 1996). Membrane extraction with a sorbent interface (MESI) coupled to GC is conceived as an exceptionally simple method for the sampling and analysis of trace compounds in the environment (Pratt and Pawliszyn, 1992a; Pratt and Pawliszyn, 1992b; Luo *et al.*, 1995; Yang *et al.*, 1994).

The MESI approach was introduced in 1992 (Pratt and Pawliszyn, 1992a; Pratt and Pawliszyn, 1992b). The original concept can be described as follows: an aqueous sample is pumped through a single hollow fibre membrane, while an inert gas flows countercurrently around the exterior of the fibre. The volatile organic compounds (VOCs) permeate from the liquid phase across the membrane and into the gas phase, where they are collected by cryofocusing and then thermally desorbed for GC analysis. MESI has been developed to allow rapid routine analysis and long-term, on-line continuous monitoring of VOCs in various environmental and industrial samples. In the MESI process, the sampling and sample preparation steps are integrated within the analytical instrument.

Generally, the apparatus for MESI (Figure 4.6) consists of four major sections (Luo *et al.*, 1995; Yang *et al.*, 1994): (i) the membrane extraction module; (ii) the cryofocusing trap and thermal desorption sorbent

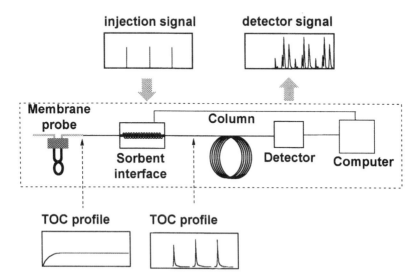

Figure 4.6 Schematic representation of membrane extraction with a solvent interface (MESI). Abbreviation: TOC, total organic compounds.

interface; (iii) the GC analysis; and (iv) the computer control and data acquisition centre.

The membrane material can be nonporous silicone rubber, polyethylene and microporous polypropylene. The advantage of using silicone is that the volatile organic analytes selectively permeate the membrane at rates that are at least a few orders of magnitude greater that the permeation of the aqueous matrix. Silicone membrane is elastic and reliable. The most practical geometric configuration for the membrane is hollow fibre. The flat sheet membrane has been used for a number of years in various separation techniques; however, the hollow fibre membrane is a more useful geometry for analytical applications because of its large surface area per volume, resulting in a more efficient extraction. The hollow fibres can also be used as probes which are inserted directly into an analyte solution. Figure 4.7 shows the geometry of the hollow fibre membrane.

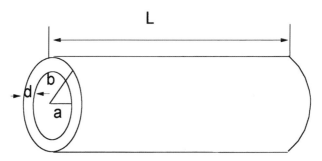

Figure 4.7 Geometry of the hollow fibre membrane. a, inner membrane radius; b, outer membrane radius; d, membrane wall thickness; L, membrane length.

The hollow fibre membrane probe can be easily installed. Normally, a piece of membrane typically 4 cm in length, is mounted onto deactivated silica tubing. The membrane is first submerged into an organic solvent, such as toluene or benzene, for half a minute; the membrane expands allowing the silica tubing to be easily slipped inside. When the membrane has been installed, the probe is exposed to air to evaporate the organic solvent. When the solvent has evaporated, the membrane shrinks to its original size and a tight seal is formed, ensuring no leakage at the junctions.

The design of the membrane module depends on different applications. Figure 4.8 shows some of the designs. The flow-over and flow-through modules are typically suitable for aqueous sample on-line monitoring. To deliver the aqueous sample, a syringe pump can be used to provide a constant sample flow rate. The headspace extraction module has the advantage of a contamination-free matrix, particularly when a dirty

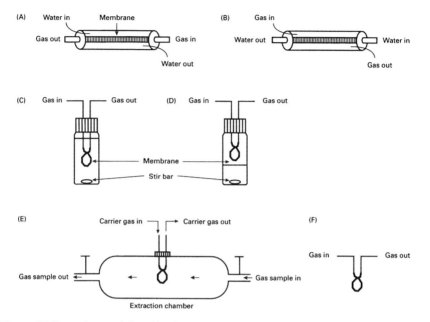

Figure 4.8 Extraction modules: A) Flow over; B) flow through; C) direct aqueous sample extraction; D) headspace extraction; E) extraction chamber; and F) air extraction.

environmental sample is analyzed. For air-monitoring, the membrane is simply exposed in air or in a chamber.

The sorbent interface contains three parts: polymer sorbent, cooling environment and pulse heating device. The polymer sorbent can be a piece of GC column or a piece of PDMS-coated, fused silica fibre or some other material. The sorbent is located in a deactivated, fused silica capillary tube. On the outside, a heating coil surrounds the tubing tightly covering the entire region where the sorbent is located. The heating coil is for the purpose of thermal desorption. The tubing with the coil is then placed in a cooling environment.

Normally, cooling temperatures between −40 and −80°C are used. When dry ice (CO_2) is used, the temperature may fall to −80°C and most VOCs can be trapped, but this method is inconvenient because dry ice is easily evaporated under ambient conditions and has to be added frequently. Another inconvenience is that dry ice is not easy to obtain or handle for field-monitoring. A semiconductive cooler has the following advantages: easy to operate; constant temperature maintainable; small geometry; and reliability. The three stage semiconductive cooler can maintain a constant temperature of −40°C with constant voltage (13 V) applied. The sorbent interface is located just before the GC injector. The

GC injector is modified, and the inlet of the column is pierced through the septum and extended to the sorbent trap.

For process automation, a computer is used to control the overall trapping process on the MESI system. To perform an analysis, the computer sends two pulses of a preset duration to the solid-state relay. The first pulse at time 0 cleans the trap. The second pulse, after a trapping period, is sent to desorb all analytes into the carrier gas for GC analysis. The second pulse also starts a computer program for real-time GC detector signal collection and display on the computer monitor. The cycle of trapping and desorbing can be repeated automatically for continuous monitoring.

MESI includes three operation modes: cryogenic trapping, multiplex and fast GC. The methods of MESI-multiplex GC and MESI fast GC have been introduced in previous papers (Luo et al., 1995; Yang and Pawliszyn, 1993). Cryogenic trapping has advantages of high sensitivity, flexible sampling time and easy operation.

The MESI cryogenic trapping approach involves on-line, cryogenic preconcentration and injection. The preconcentration step allows high sample throughput and enhances system sensitivity. Using a cryogenic coolant increases the absorption capacity of the sorbent trap and focuses concentration bands to improve the efficiency of chromatographic separation. In this operation mode, the analytes are stripped off from the inner surface of the membrane, transferred with the carrier gas and accumulated at the sorbent for a certain period of time. After a period of trapping time, a heating pulse is sent to desorb all compounds from the trap. Since the pulse is short, 2 s for example, a sharp concentration band is generated at the inlet of the GC column. The cycle of trapping and heating can be repeated for continuous monitoring. The sensitivity of MESI in the trapping method is directly related to the trapping time, which is limited by the breakthrough time of the analytes at the sorbent interface. A longer trapping time results in a greater accumulation of analyte and larger detector response, and hence higher sensitivity.

4.3.2 Theoretical aspects of the membrane extraction process

Based on the extraction modes, the extraction can be direct aqueous, gaseous and headspace. For direct aqueous sample extraction, the concentration distribution is depicted in Figure 4.9. Analyte transport in the MESI system is divided into five steps (Luo et al., 1997): (i) convection and diffusion through the sample to the outer surface of the membrane; (ii) partitioning between the sample and membrane at its outer surface; (iii) diffusion through the membrane; (iv) partitioning between the membrane and mobile phase at the inner surface of the

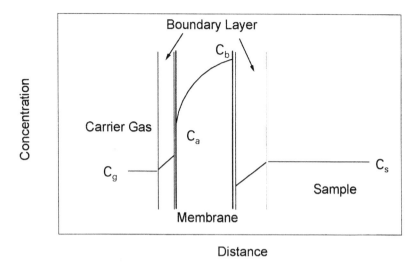

Figure 4.9 Schematic representation of concentration distribution in aqueous sample extraction. Abbreviations: C_a, concentration at inner membrane surface; C_b, concentration at outer membrane surface; C_g, concentration in the stripping gas; C_s, sample concentration.

membrane; and (v) diffusion and convection of the analyte into the mobile phase, which flows out of the membrane.

The diffusion in the membrane undergoes two processes: non-steady-state and steady-state. A non-steady-state process refers to the formation of a concentration gradient in the membrane, and in this process the extraction rate is changed with time. A steady-state process means that there is a constant concentration gradient in the membrane and the extraction rate is not varied with time. Figure 4.10 indicates the two processes.

The extraction rate, G_e, under steady-state can be expressed as:

$$G_e = \frac{AD_m K_{ms} C_s}{k'_1 + k_1 + \ln b/a} \qquad (4.9)$$

where, a and b are the radius from the membrane axis to the membrane inner and outer surface, respectively, C_s is the concentration of analyte in the bulk sample, K_{ms} is the distribution constant of the analyte between sample and membrane, and D_m is the diffusion coefficient of the analyte in the membrane. A is the surface area of the membrane. The parameter k'_1 is a measure of the resistance to mass transfer at the outer surface of the membrane, and the parameter k_1 is a measure of resistance to mass transfer at the inner surface of the membrane; k'_1 and k_1 can be reduced

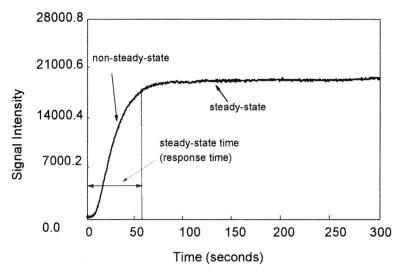

Figure 4.10 Extraction time profile of benzene.

by using good sample agitation and high stripping gas flow rate, respectively. It can be seen in Equation 4.9 that the extraction rate is proportional to sample concentration; this is the basis of quantitation in MESI. It can also be seen that the impact factors to the extraction rate are distribution constant, diffusion coefficient, membrane surface area, carrier gas flow rate and agitation.

The steady-state time, which is defined as the time of a permeation reaching 90% of steady-state, can be expressed as:

$$t_{90\%} = \frac{a^2}{2D_m} \frac{[1+(b/a)^2]\ln(b/a)-(b/a)^2+1+[(b/a)^2-1-2\ln(b/a)]k_1}{k_1'+k_1+\ln(b/a)}$$
$$+ \frac{-[(b/a)^2-1-2(b/a)^2\ln(b/a)]k_1' + 2[(b/a)^2-1]k_1'k_1}{k_1'+k_1+\ln(b/a)}$$

(4.10)

The above equation indicates that the steady-state time depends on: the diffusion coefficient in the membrane; the membrane geometry; and agitation on both sides of the membrane.

For gaseous phase extraction, the boundary layers are not a significant factor; this is due to the fact that the analyte has a large diffusion coefficient in gaseous phase. Generally, diffusion coefficients in gases are about $0.1\,\text{cm}^2/\text{s}$ and $10^{-5}\,\text{cm}^2/\text{s}$ in liquids. Therefore, the diffusion coefficient of analytes in gases is five to six orders greater in magnitude than in liquids. For most VOCs, if the extraction is in the gas flow stream, the boundary layer effect is negligible. However, if the extraction is in static-state, the boundary layer may be significant for some VOCs which have relatively small diffusion coefficients. In this case, the boundary layer should be considered, as above. The extraction rate in gaseous phase without considering the boundary layer effect at the steady-state can be described as:

$$G_s = C_s A D_m K_{ms} N \qquad (4.11)$$

where, N is a constant which relates to the flow rate and concentration in the stripping gas.

For headspace extraction of MESI, analytes undergo a series of transport processes from water to the gas phase and from the gas phase to the membrane. The extraction rate depends not only on sample agitation but also on the diffusion coefficient of the analyte and the distribution constant between the sample matrix and gaseous phase. Details of the discussion can be referred to in the paper by Yang and co-workers (1996).

4.3.3 *Practical aspects of MESI analysis*

In MESI, some factors significantly affect the level of extraction and trapping efficiency (Luo *et al.*, 1997a; Luo *et al.*, 1998). A proper optimization ensures high extraction and trapping efficiency, which means a good sensitivity, good quantitation and practical monitoring.

4.3.3.1 *Optimization of trapping*

A sorbent trap includes the sorbent, cooling environment and pulse heating device. Trapping is impacted by breakthrough and, when a breakthrough occurs, a serious quantitation problem arises; therefore, in MESI trapping, breakthrough should ultimately be avoided. To achieve trapping without breakthrough, a proper sorbent should be carefully selected. Different kinds of sorbent are available, which can be polar or nonpolar, liquid or solid coating polymer. Selection of the coating is based primarily on the polarity and volatility characteristics of the analyte. In addition, a proper sorbent volume should be considered. In general, a large sorbent volume results in a large trapping capacity, and hence a high sensitivity. However, because the thermal capacity of the sorbent trap determines the width of the desorption pulse, a large sorbent volume needs a long desorption time and hence a wide injection band or a serious carry-over. Thus, a smaller sorbent trap is better for rapid desorption, and the sorbent capacity should be increased without significant increase of the thermal mass of the sorbent.

In MESI, sensitivity is proportional to trapping time; a good sensitivity can be obtained under conditions of low trapping temperature and long trapping time. However, the trapping time is limited by breakthrough, so a proper trapping time should be determined in a practical application. Generally, for a highly concentrated sample, a short trapping time can satisfy the needs of sensitivity, and for low concentration, a long trapping time should be applied. A maximum trapping time should be less than the breakthrough time, which should be carefully checked. For thermal desorption, heating intensity and pulse width depend on the trapping temperature and the stability of the sorbent. A low heating intensity results in serious carry-over and overheating causes sorbent degradation. The pulse heating width also relates to carry-over. A pulse heating of too short a duration normally results in significant carry-over, and a long pulse heating affects the lifetime of the sorbent. In practice, for a 1 cm long PDMS fibre (100 mm) under $-40°C$ trapping, a pulse of $280°C$ and 2 s duration can be used for benzene, toluene, ethylbenzene, xylene (BTEX) analysis and no carry-over is observed.

4.3.3.2 Effect of significant factors on membrane extraction
Efforts to improve the MESI procedure are directed at obtaining high extraction efficiency and short response time. The response time can be expressed by steady-state time as defined by Equation 4.10.

Membrane selection depends on characteristics such as selectivity, and mechanical or chemical durability. The chemical nature of the target analyte determines the type of membrane used. In this selection, the distribution constant and diffusion coefficient of the compound are the two most important parameters. The distribution constant determines the sensitivity of the extraction, and the diffusion coefficient determines the mass flux across the membrane. A large distribution constant indicates that the analyte has a good affinity to the membrane. However, a large distribution constant does not ensure high sensitivity; only with a high diffusion coefficient can a good sensitivity be ascertained.

Extraction efficiency relates to the surface area of the membrane; a large surface results in a high extraction rate. For the hollow fibre membrane, a large surface area can be obtained by extending the membrane length. However, an over long membrane does not mean a high extraction efficiency if a low stripping gas flow rate is applied; and, in many cases, a lower extraction efficiency may be observed because of the absorption of the extended membrane section. When a high stripping gas flow rate is used, breakthrough becomes a problem. An over long membrane is not good for practical operation; it needs to be coiled to fit into a sample vial, which is inconvenient. In practice, a 4 cm long membrane with a 2–5 ml/min stripping gas flow rate can be chosen.

The extraction rate increases with a decrease in the thickness of the membrane wall. The reason for this increase is apparent considering the concentration gradient along the membrane thickness (Figure 4.9). A thinner wall thickness leads to a higher concentration at the inner surface, which means higher flux of analyte into the carrier gas, and hence a higher overall extraction rate. A thinner wall membrane also gives rise to a short response time.

The extraction rate depends on the stripping gas flow rate. At a low flow rate, the stripping gas has a relatively long time in contact with the inner surface of the membrane. The analyte easily reaches partition equilibrium between the stripping gas and the inner surface of the membrane. In this case, the stripping gas can obtain a relatively high concentration but, because of the low flow rate, a relatively small amount of analyte is stripped off from the inner surface of the membrane per unit time. Thus, the overall extraction rate is low. At a higher flow rate, the carrier gas either does not reach partition equilibrium or reaches partition equilibrium near the membrane exit. Although the carrier gas has a relatively low concentration compared to that in a lower flow rate, the overall extraction rate is higher because of the higher flow rate. When a high flow rate is used, care should be taken to ensure no breakthrough.

The extraction temperature significantly affects the extraction efficiency. This effect is due to the changes of distribution constant and diffusion coefficient. The effect on distribution constant can be described as (Crank and Park, 1968):

$$K = K_0 \exp\left[\frac{\Delta H}{R}\left(\frac{1}{T} - \frac{1}{T_0}\right)\right] \quad (4.12)$$

where, T is the temperature in degrees Kelvin, K_0 is the distribution constant at temperature T_0, ΔH is the change in enthalpy when the analyte passes from air into the membrane, and R is the gas constant. ΔH is considered constant for the ambient temperature range, and is close to the value of ΔH^v, the enthalpy change of vaporization of the pure analyte. From Equation 4.12, it can be seen that the K value decreases with an increase in temperature. The effect on the diffusion coefficient can be expressed as:

$$D = D_0 \exp(-E_d/RT) \quad (4.13)$$

where, D_0 is a pre-exponential factor, and E_d is the apparent activation energy for diffusion. It can be seen that the impact of temperature on these two parameters is contrary. Normally, for air extraction, the extraction efficiency decreases with an increase in temperature, and for direct aqueous sample extraction, the extraction efficiency increases with an increase in temperature. For headspace extraction, since an increase in temperature is good for release of analytes from the matrix, an increased extraction efficiency is obtained.

Agitation can greatly improve the extraction efficiency by improving mass transfer in the sample matrix and reducing the boundary layer, and essentially reducing response time. Magnetic stirring is most commonly used in MESI for both direct aqueous sampling and headspace extraction. Care must be taken when using this technique to ensure the temperature is not changed. Magnetic stirring is efficient when fast stirring speeds are used. Sonication is another agitation approach, although it is sometimes not very effective.

A small headspace can increase the extraction rate from aqueous samples. This is because a compressible headspace allows the carrier gas to penetrate the membrane wall and form bubbles on the outer surface of the membrane. Although the analyte still needs to diffuse through water to the gas bubbles and then be extracted, mass transfer is enhanced because the small gas bubbles have a large surface area, the molecules of analyte can easily diffuse through these gas bubbles and, in general, compounds have larger distribution constants between membrane/air than between membrane/water. With a headspace, concentration in the aqueous solution drops somewhat because the analytes distribute into the headspace, but overall the extraction rate is increased if the headspace is small. The extraction amount is reduced when the headspace volume is further increased. This reduction occurs because more molecules of analyte distribute into the headspace. The concentration in the solution is decreased significantly. Therefore, when headspace is applied to aid direct aqueous sample extraction, the headspace should be kept as small as possible.

If MESI is applied directly to an effluent stream or biological system rich in organic compounds, significant interference could cause high background noise or affect the parameters that govern response time and extraction rate in a nonreproducible manner. Humic or other materials in a sample could foul the membrane. In that situation, a headspace approach is more suitable.

In MESI, the extraction rate is pressure-dependent; a pressure difference between the outside and inside of the membrane changes the diffusion coefficient in the membrane. This effect results in a change in extraction rate; however, this change is not significant in ambient conditions where the pressure may change from place to place. When the extraction is in a high pressure system, membrane squeezing should be considered, which may cause clogging and stop the stripping gas flow.

Figure 4.11 demonstrates the monitoring of fermentation with MESI. The monitoring was performed at 1, 10 and 24 h after a flour fermentation. In monitoring, the membrane probe was positioned in the headspace of the broth, the extraction (sampling) time was 3 min and the sorbent trapping was under $-40°C$. Note that the products and concentration varied with time. From this application, it can be seen that the method of microextraction with a sorbent interface for on-line monitoring combines sampling, preconcentration and injection in one step; no organic solvent was used and the operation was simple.

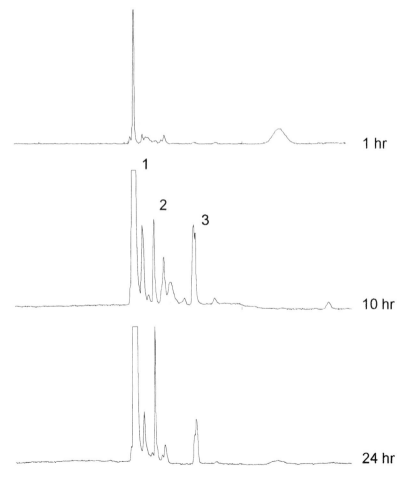

Figure 4.11 Monitoring of a fermentation using membrane extraction with a sorbent interface (MESI): 1, ethanol; 2, acetic acid; 3, acetoin.

Acknowledgements

The authors thank the Natural Sciences and Engineering Research Council of Canada, the Dow Chemical Company, Chrompack, Varian and Supelco for financial support.

References

Bier, M.E. and Cooks, R.G. (1987) Membrane interface for selective introduction of volatile compounds directly into the ionization chamber of a mass spectrometer. *Anal. Chem.*, **59** 597-601.

Crank, J. and Park, G.S. (1968) *Diffusion in Polymers*, Academic Press, London, UK.

Gorecki, T. and Pawliszyn, J. (1995) Sample introduction approaches for solid phase microextraction/rapid GC. *Anal. Chem.*, **67** 3265-74.

Jönsson, J.Å. and Mathiasson, L. (1992) Supported liquid membrane techniques for sample preparation and enrichment in environmental and biological analysis. *Trends Anal. Chem.*, **11** 106-14.

LapPack, M.A., Tou, J.C. and Enke, C.G. (1990) Membrane mass spectrometry for the direct trace analysis of volatile organic compounds in air and water. *Anal. Chem.*, **62** 1265-71.

Lauritsen, F.R. and Gylling, S. (1995) On-line monitoring of biological reactions at low parts-per-trillion levels by membrane inlet mass spectrometry. *Anal. Chem.*, **67** 1418-20.

Lindegård, B., Bjork, H., Jönsson, J.Å., Mathiasson, L. and Olsson, A. (1994) Automated column liquid chromatographic determination of a basic drug in blood plasma using the supported liquid membrane technique for sample pretreatment. *Anal. Chem.*, **66** 4490-97.

Lotiaho, T., Lauritsen, F.R., Choudhury, T.K., Cooks, R.G. and Tsao, G.T. (1991) Membrane introduction mass spectrometry. *Anal. Chem.*, **63** 875A-83A.

Luo, Y.Z., Yang, M.J. and Pawliszyn, J. (1995) Membrane extraction combined with a sorbent coated fibre interface for capillary gas chromatography. *J. High Resol. Chromatogr.*, **18** 727-31.

Luo, Y.Z., Adams, M. and Pawliszyn, J. (1997) Aqueous sample direct extraction and analysis by membrane extraction with a sorbent interface analysis. *Analyst*, **122** 1461-69.

Luo, Y.Z., Adams, M. and Pawliszyn, J. (1998) Kinetic study of membrane extraction with a sorbent interface for air. *Anal. Chem.*, **70** 248-54.

Martos, P., Saraullo, A. and Pawliszyn, J. (1997) Estimation of air/coating distribution coefficients for solid phase microextraction using retention indexes from linear temperature-programmed capillary gas chromatography: application to the sampling and analysis of total petroleum hydrocarbons in air. *Anal. Chem.*, **69** 402-408.

Martos, P. and Pawliszyn, J. (1997) Calibration of solid phase microextraction for air analyses based on physical chemical properties of the coating. *Anal. Chem.*, **69** 206-15.

Pawliszyn, J. (1997) *Solid Phase Microextraction. Theory and Practice*, Wiley-VCH, New York, USA.

Pratt, K.F. and Pawliszyn, J. (1992a) Gas extraction kinetics of volatile organic species from water with a hollow fibre membrane. *Anal. Chem.*, **64** 2101-106.

Pratt, K.F. and Pawliszyn, J. (1992b) Water monitoring system based on gas extraction with a single hollow fibre membrane and gas chromatographic cryotrapping. *Anal. Chem.*, **64** 2107-110.

Schwarzenbach, R., Gschwend, P. and Imboden, D. (1993) *Environmental Organic Chemistry*, John Wiley and Sons Inc., New York, USA, pp. 109-23.

Thordarson, E., Pálmarsdóttir, S., Mathiasson, L. and Jönsson, J.Å. (1996) Sample preparation using a miniaturized supported liquid membrane device connected on-line to packed capillary liquid chromatography. *Anal. Chem.*, **68** 2559-63.

Virkki, V.T., Ketola, R.A., Ojala, M., Kotiaho, T., Komppa, V., Grove, A. and Faccchetti, S. (1995) On-site environmental analysis by membrane inlet mass spectrometry. *Anal. Chem.*, **67** 1421-25.

Yang, M.J. and Pawliszyn, J. (1993) Multiplex gas chromatography with a hollow fiber membrane interface for determination of trace volatile organic compounds in aqueous samples. *Anal. Chem.*, **65** 1758-63.

Yang, M.J., Harms, S., Luo, Y.Z. and Pawliszyn, J. (1994) Membrane extraction with a sorbent interface for capillary gas chromatography. *Anal. Chem.*, **66** 1339-46.

Yang, M.J., Adams, M. and Pawliszyn, J. (1996) Kinetic model of membrane extraction with a sorbent interface. *Anal. Chem.*, **68** 2782-89.

Zhang, Z. and Pawliszyn, J. (1995) Quantitative extraction using an internally cooled solid phase microextraction device. *Anal. Chem.*, **67** 34-43.

Zhang, Z. and Pawliszyn, J. (1996) Studying activity coefficients of probe solutes in selected liquid polymer coatings using solid phase microextraction. *Phys. Chem.*, **100** 17648-54.

5 Supercritical fluid extraction in organic analysis
Hans-Gerd Janssen and Xianwen Lou

5.1 Introduction

The extraction process is one of the most important steps in pretreatment for solid and liquid samples. It is therefore not surprising that a wide range of techniques has been developed for the extraction of these samples. The most widely-used method for the extraction of solid samples is Soxhlet extraction. In recent years, various new techniques for sample preparation, such as Microwave-Assisted Extraction (MAE) (Ganzler et al., 1986; Lopez-Avilla et al., 1995), Supercritical Fluid Extraction (SFE) (Hawthorne et al., 1990; Chester et al., 1992; Lee and Markides, 1990), automated Soxhlet and Accelerated Solvent Extraction (ASE) (Richter et al., 1996), have been developed as alternatives to Soxhlet. Compared to Soxhlet extraction, each of the new techniques reduces the amount of solvent required and/or shortens the sample preparation time. The time- and solvent-consuming nature of Soxhlet extraction is generally imputed to the slow diffusion of the analytes from the sample matrix into the extraction fluid and/or the slow desorption of the components from the sample matrix.

By the introduction of microwave or sonication or by extracting the components at elevated temperatures, the rates of diffusion and desorption can be significantly increased, resulting in shorter extraction times. Similar advantages can also be obtained by using a supercritical fluid as the extractant. Extractions using supercritical carbon dioxide as the extraction fluid can have the additional advantage that the solvent used for extraction is environment- and analyst-friendly, and that on-line combination of the sample preparation step with chromatographic analysis, if desired, is relatively simple. Further advantages of SFE using CO_2 as the extractant include: the chemical inertness of CO_2; the high purity and low price; the nonflammable nature; the easily accessible critical parameters; and the ease with which the solvent can be removed after extraction. It is for these reasons that CO_2 is by far the most widely-used fluid in SFE.

Historically, the development of analytical SFE is associated with the introduction of Supercritical Fluid Chromatography (SFC). SFC was introduced in the early 1980s as a technique combining the advantages of gas chromatography (GC) and high-performance liquid chromatography (HPLC). High molecular weight solutes could be chromatographed at a low temperature using a supercritical eluent as the mobile phase. When using pure CO_2 as the mobile phase, the universal and very sensitive GC flame ionisation detector could be used for solutes that normally required separation by liquid chromatography (LC) as a result of their low vapour pressure. SFC was therefore seen as a highly promising technique for the analysis of high molecular weight and/or thermally unstable analytes for which neither GC nor LC could be used. Indeed a (limited) number of chromatographic applications is nowadays routinely performed using SFC as the chromatographic method. In addition to this, SFC has had an important spin-off in that it has introduced supercritical fluids into the analytical laboratory.

Although the ability of supercritical fluids to dissolve solids had been known for many years (Stahl et al., 1978; Francis, 1954), it was only after seeing the possibilities of SFC that analytical SFE was rapidly adopted for use as an extraction technique. What struck the chromatographer in SFC was that retention and selectivity could be varied over an extremely wide range simply by adjusting two instrumental parameters, temperature and pressure. The need for selective methods of sample preparation in combination with the observation from SFC that varying the selectivity of a supercritical fluid required merely a pressure change, inspired chromatographers to use supercritical fluids for sample preparation. These initial SFE experiments could be readily performed using SFC instrumentation. Soon afterwards, SFE started to overtake SFC in popularity. In the mid-to-late 1990s, SFC has clearly become an established technique that holds a secure, but very small and no longer increasing, niche position in the analytical laboratory next to the much larger techniques, LC and GC. Progress in the use of supercritical fluids for extractions has clearly slowed down, but new areas of application are still emerging.

SFE is still a developing technique. Initial enthusiasm, or even euphoria, about SFE gave way to a more realistic view of its possibilities and limitations. SFE is clearly a technique that has unique possibilities but, unfortunately, it also presents some unique problems. Euphoria often ends in disappointment. In part, this is already the case with SFE. The technique is powerful, but not always easy to use and optimise. Many parameters affect the final result of an SFE experiment; therefore, numerous possibilities for fine-tuning the extraction behaviour are presented. Unfortunately, the drawback of this is that many pitfalls are also encountered. However, experienced analysts with a good knowledge of SFE, can use the strengths of this extraction method to their advantage, without experiencing the pitfalls that SFE can create.

This chapter will describe the principles of SFE, starting with a brief description of the properties of supercritical fluids. The instrumental requirements for SFE will be discussed in detail. As method development in SFE and, related to this, the optimisation of experimental conditions is generally considered to be the most difficult step, this subject will be treated in detail. Attention will be paid to each of the parameters that affect the performance of SFE experiments. Finally, a general overview of the most important areas of application will be presented. The strengths and weaknesses of SFE in these areas of application will be identified, and the consequences of such for future directions and new developments will be addressed in brief.

5.2 Properties of supercritical fluids

A fluid is said to be in its supercritical state when both its temperature and pressure are above their respective critical values. This definition can be visualised by reference to the pressure/temperature diagram of CO_2 presented in Figure 5.1. Supercritical fluids can be seen as intermediates

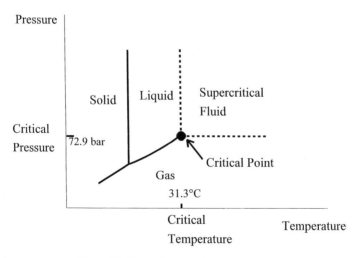

Figure 5.1 Phase diagram of carbon dioxide.

between gases and liquids. The distance between the molecules can be varied continuously between those of a liquid, where the individual molecules are very close to each other and strong forces act between them, to those found in a gas. In the latter case, the individual molecules move at great distances and forces between the molecules are weak. Under conditions where the density of the fluid is large, liquid-like conditions prevail and the supercritical fluid is a good solvent. At the molecular level, strong forces are exerted between the supercritical solvent molecules and the molecules that are to be extracted. The supercritical extractant now solubilizes the target molecules, resulting in efficient extraction conditions. Under higher temperature and/or lower pressure conditions, the fluid behaves in a more gas-like manner and is therefore a much weaker solvent. Under these conditions, mutual forces between the molecules are weak, making the supercritical extractant a weak solvent.

Supercritical fluids offer, at least in principle, liquid-like solvating powers with more gas-like transport properties. This combination of properties makes supercritical fluids attractive both for chromatography and extraction. Physical properties of supercritical fluids which prove to be important in extraction are the density ('related to the molecular closeness') and the viscosity and diffusivity of the fluid. The first of these parameters, density, relates to the solvent strength, whereas the second, diffusivity, gives information on the rate of equilibration. Table 5.1

Table 5.1 Approximate values of densities, viscosities and diffusion coefficients of gases, supercritical fluids and liquids

Fluid	Density g/cm^3	Viscosity $g/cm \cdot s^{-1}$	Diffusion coefficient cm^2/s
Gas	$0.6–2 \cdot 10^{-3}$	$1–3 \cdot 10^{-4}$	$0.1–1$
Supercritical fluid	$0.2–0.9$	$1–3 \cdot 10^{-3}$	$0.1–5 \cdot 10^{-4}$
Liquid	$0.6–1.6$	$0.2–3 \cdot 10^{-2}$	$0.2–3 \cdot 10^{-5}$

(Randall, 1982).

presents a comparison of these properties for gases, supercritical fluids and liquids.

The combination of liquid-like solvent strength and gas-like transport properties of supercritical fluids makes SFE an important alternative to traditional Soxhlet extraction. Ideally, SFE offers extraction strengths comparable to that of conventional liquid extractions, while simultaneously providing gas-like kinetics, i.e. equilibrium conditions are established rapidly. The extraction strength of a supercritical fluid is determined by the density of the fluid. This parameter, in turn, is determined by the pressure and temperature of the supercritical fluid. An important advantage of supercritical fluids is that their solvent power can be adjusted through an instrumental parameter, pressure. This feature not only allows selective extraction to be obtained but also allows for rapid preconcentration of the extract after extraction, simply by releasing pressure. In this way, the target solutes can be obtained free from any contaminating solvent. A clear example of the dramatic influence of pressure on the extraction behaviour of CO_2 is presented in Figure 5.2.

This figure shows the supercritical fluid extraction of cold-pressed grapefruit oil using pure CO_2 at different densities. At low densities, almost no components are extracted, whereas large amounts are extracted at higher densities.

The description of the effect of pressure on the solvent strength in SFE, as presented above, is highly qualitative and superficial. A more accurate model was presented by Giddings and co-workers (1968). According to these authors, the behaviour of a supercritical fluid in an extraction experiment (or in chromatography) can be modelled as consisting of a 'state effect', described by physical properties such as density, and a 'chemical effect', which takes into account specific interactions between the solvent and solute molecules. In principle, this introduces the possibility of using specific supercritical solvents that selectively enhance the solubility of a particular class of analytes during SFE. In practice,

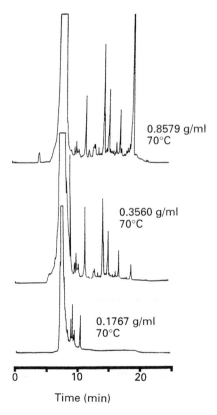

Figure 5.2 SFE-SFC of cold-pressed grapefruit oil at different extraction densities. The sample was extracted with CO_2 for 12 min at 70°C. (Andersen *et al.*, 1989; reproduced with permission from the *Journal of Chromatographic Science*).

however, this is hardly ever applied. Most SFE extractions are performed using either pure or modified CO_2 as the mobile phase; hence, the possibility of exploiting the 'chemical effect' is limited.

An overview of a number of other possible solvents for use in SFE is given in Table 5.2. These solvents, and other potential solvents not listed here, are seldom used for reasons of safety, price, impractical critical conditions, corrosivity, stability, purity, etc. For example, Freons have been shown to yield better recoveries for polycyclic aromatic hydrocarbons (PAHs) and polychlorinated biphenyls (PCBs) in the SFE of environmental samples but their adverse effects on the atmospheric ozone layer make them unacceptable for routine use (Hawthorne et al., 1992; Howard et al., 1993; Chambers et al., 1979). Nitrous oxide has a permanent dipole moment and is, therefore, a better solvent for many solutes than CO_2 (Hawthorne et al., 1990). However, it is a strong oxidising agent and can present a serious safety hazard. Supercritical methanol is an excellent solvent but is a liquid in ambient conditions, which complicates sample collection and concentration after extraction. Supercritical ammonia would be very attractive from the point of view of solvent strength, but it is chemically reactive and is likely to be too dangerous for routine use (Hawthorne, 1990).

Table 5.2 Characteristics of selected supercritical fluids

Fluid	Tc °C	Pc atm	Dipole moment Debyes
CO_2	31.3	72.9	0.00
N_2O	36.5	72.5	0.17
NH_3	132.5	112.5	1.47
MeOH	240.0	78.5	1.70
Xe	16.6	58.4	0.00
CCl_2F_2	111.8	40.7	0.17
$CClF_3$	28.8	38.2	0.50
Ethane	32.2	48.2	0.00
Ethylene	9.9	50.5	0.00

Abbreviations: Tc, critical temperature; Pc, critical pressure. (Weast, 1984).

The influence of pressure and temperature on the extraction behaviour of solutes in SFE is, generally, not easy to understand. This, despite considerable efforts to improve the understanding of the fundamental mechanisms underlying SFE (Chester et al., 1992; Lee and Markides, 1990; Gere et al., 1993; Lou et al., 1995a; Bartle et al., 1989; King, 1989; McNally and Wheeler, 1988). More detailed information on this subject will be presented in Section 5.5, where method development in SFE is discussed. For the present, it should be remarked that even the prediction of solubilities in supercritical fluids is already complicated. Prediction of the entire SFE process, which involves aspects such as solubility in the extractant and interaction of the target solute with the matrix in addition to kinetic parameters, is as yet only possible for simple systems, such as the extraction of polymer additives from well-defined, ground polymer particles (Bartle *et al.*, 1989).

An oversimplified, yet useful, model for understanding the influence of pressure on solubility in a supercritical fluid assumes that the solubility of a component in a supercritical fluid is controlled by two parameters: the vapour pressure of the solute and its interaction with the supercritical fluid. The latter parameter is closely related to density. With this model in mind, the generally observed trends in extraction yields with changes in pressure or temperature can be understood. Changes in either pressure or temperature result in changes in density. An increase in pressure at constant temperature results in improved extraction yields because of the better solubility at increased fluid densities. The effect of temperature variations is more complicated. If temperature is increased at constant density, the solubility in the supercritical fluid will generally increase because of the increased vapour pressure of the solute. Increasing temperature at constant pressure, on the other hand, results in behaviour that is far more complicated. Very often the solubility will decrease as a result of the lower density, although occasionally an increase in solubility can be observed when increasing temperature at constant pressure.

Although oversimplified, the discussion presented above has proved to be useful for daily SFE practice. It should be emphasised, however, that the model presented here is not only oversimplified but in some cases even incorrect. An example of the complicated solubility behaviour of organic compounds in a supercritical fluid is presented in Figure 5.3.

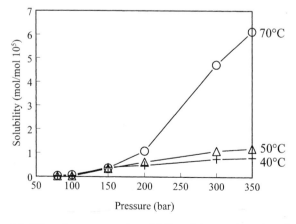

Figure 5.3 Effect of pressure on the solubility of chrysene in carbon dioxide.

This figure shows the solubility of chrysene in supercritical CO_2 as a function of pressure at different temperatures (Lou *et al.*, 1997). The information on solubilities in supercritical fluids published in the literature is sometimes contradictory. For example, Zhao and co-workers (1995) found that in the near-critical region the solubilities of some PAHs in supercritical CO_2 decreased significantly when increasing temperature at constant pressure. Apparently, the solubility here is governed by the interactions of the solutes with the fluid and vapour pressure is of minor importance. Miller and Hawthorne (1995), on the contrary, reported a continuous increase in the solubility of these solutes when increasing temperature at a constant pressure despite the decrease in CO_2 density, indicating that vapour pressure is the predominant parameter. We found that at pressures above 200 bar the solubility of anthracene increased continuously with increasing temperature, whereas at 100 bar it first decreased and than increased with temperature (Lou *et al.*, 1997).

These observations clearly show that more sophisticated models, such as the solubility parameter concept (Giddings *et al.*, 1968), are required to describe solubility in supercritical fluids. Using the solubility parameter theory, the apparently conflicting observations on the solubilities of PAHs in supercritical CO_2 referred to above can indeed be explained. Unfortunately, however, the solubility parameter theory can only provide a qualitative explanation of the observed effects of temperature and pressure conditions on solubilities in supercritical fluids. For more quantitative explanations or even predictions, more detailed physicochemical parameters would be required, that are not generally available.

5.3 Modifiers or co-solvents

Carbon dioxide has a large number of attractive features for use as an extraction fluid in SFE. A distinct disadvantage of CO_2, however, is its relatively nonpolar nature. Analogous to the situation in SFC, it frequently becomes necessary to add a modifier to the supercritical fluid to enhance the elution/extraction strength of the fluid in the extraction of polar and/or high molecular weight solutes. The modifiers, in SFE also called co-solvents, entrainers or moderators, are usually polar organic solvents that are added to the source of compressed fluid before the pump, added to the extraction gas after it is compressed or, alternatively, spiked onto the matrix that is to be extracted. In the latter case, the modifier is gradually eluted out of the system. Popular modifiers are the lower alcohols, ethylacetate, acetone, hexane, toluene, formic acid, water, etc. The mechanism of modifier effects has received a great deal of attention in SFC (Janssen *et al.*, 1991). Much of the knowledge gained is also applicable to SFE. Retention in SFC is governed by a combination of the properties of both the mobile and the stationary phase in relation to those of the solute. In SFE, a similar situation occurs; here, the matrix replaces the stationary phase. The addition of a modifier to the system can induce changes in the nature both of the extractant and the matrix. Hence, the effects of adding a modifier can be divided into an extraction-fluid modification effect and a matrix effect. The possible effects of the modifier in the SFE system consisting of extraction fluid and matrix are schematically depicted in Figure 5.4.

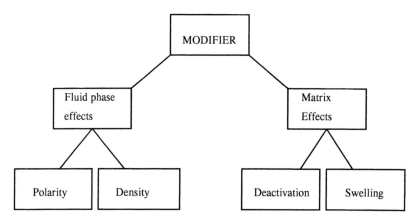

Figure 5.4 Schematic illustration of the effects of modifiers in supercritical fluid extraction (SFE). For a discussion of the various mechanisms see Section 5.3.

Properties of the extractant that are liable to change upon the addition of modifiers include the density, and the nature and extent of physicochemical interactions between the solute and the modified extractant. In/on the matrix, the use of modifiers can lead to displacement of solute molecules from active adsorptive sites. Additional effects of modifiers might include physical swelling of the matrix resulting in enhanced mass transfer in the solid matrix particles. As stated previously, the addition of a modifier to the (CO_2) extractant phase may lead to an alteration in the density of the fluid compared to pure CO_2 at identical conditions of pressure and temperature. For mixed extraction fluids, a higher density is anticipated with the commonly used modifiers because the critical parameters of the mixed fluid are higher than those of pure CO_2. The use of a modified extraction phase is, therefore, a possible means of increasing density when the maximum pressure of the system is approached. In addition to changing the density of the extraction fluid, polar modifiers, such as the lower alcohols, also increase the polarity of the fluid. This can significantly enhance the solubility of polar solutes in the originally nonpolar CO_2. This improvement in solubility is the result of specific interactions between solute molecules and modifier molecules in the extractant phase. These interactions can include dipolar, hydrogen-bonding and dispersive interactions as well as ion-pairing mechanisms.

The solubility of various classes of solutes in CO_2 modified with polar and nonpolar solvents has been the subject of a detailed study (Dobbs *et al.*, 1987). The modifier-induced enhancement in solubility could be understood qualitatively using dispersion, orientation and acid-base solubility parameters. For example, the enhancement in solubility for benzoic acid in CO_2/modifier mixtures was significantly greater with methanol as the modifier than with n-octane. In contrast, the solubility of hexamethylbenzene was twice as high in CO_2/n-octane compared with CO_2/methanol. All solubilities were measured at the same density. The greatest enhancement in solubility was obtained when the nature of the modifier matched that of the solute. In addition to the above-mentioned enhancements in solubility due to selective interactions, ionic species or highly polar solutes can be extracted by using more exotic forms of 'solubility-tuning' through the addition of derivatization, ion-pairing or complexation agents (Fields, 1997). Irrespective of which approach is used, the use of modifiers virtually always complicates the use of SFE as a sample pretreatment technique. The critical parameters of the mixture change, phase separation can occur, the complexity of the instrumentation required is increased, selectivity is generally reduced and isolation of the components of interest from the extractant is more difficult. Two examples of the use of modified extraction fluids are presented in Figures 5.5 and 5.6. Figure 5.5 shows the effect of the addition of a modifier on the extraction of poppy straw. This is an example of an extraction where selectivity can be introduced through the use of a modifier. The GC analysis of the SFE extract obtained using pure CO_2 shows large peaks for the terpenes from this natural product (Figure 5.5A). Figure 5.5B shows the subsequent extraction of the same sample, now using 5% methanol as a modifier. The GC chromatogram of this extract shows large peaks for alkaloids, polar species that could not be extracted using neat CO_2.

An example of the use of ion-pairing reagents for the extraction of ionic compounds is presented in Figure 5.6. This figure shows the SFC chromatogram obtained after ion-pairing SFE of berberine and palmatine, two alkaloids from *Phellondendri cortex*, a herb prescribed in many Japanese and Chinese traditional medicines. The ion-pairing SFE method uses dioctyl sodium sulphosuccinate (DSS) to neutralise the components of interest prior to extraction. Formation of an ion-pair results in a neutral complex. In this way, the polarity of the ionic compound is reduced and its solubility in the CO_2-methanol mixture is greatly increased.

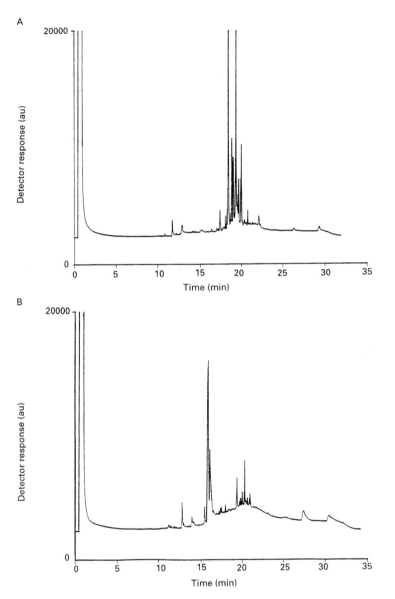

Figure 5.5 Extraction of polar alkaloids from poppy straw. A) Extraction using pure CO_2 at temperature 40°C, pressure 200 bar, extraction time 30 min, and analysis by gas chromatography (GC). B) Extraction fluid 20% ethanol in CO_2, and other conditions as for example A. Abbreviation: au, arbitrary units.

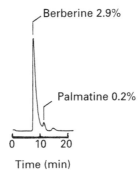

Figure 5.6 Ion-pairing SFC chromatogram of an ion-pair SFE extract from *Phellodendri cortex*. SFE conditions: temperature 60°C; pressure 200 bar; flow rate 4.0 ml/min; extraction fluid CO_2/MeOH, 90/10 v/v, containing 100 mM dioctyl sodium sulphosuccinate (DSS); and extraction time 10 min. SFC conditions: temperature 60°C; pressure 200 bar; flow rate 4.0 ml/min; mobile phase CO_2/MeOH; 85/15 v/v, containing 100 mM DSS. Chromatographic column, LiChrosorb Si60 (150 × 4.6 mm ID, 10 μm); UV detection at 345 nm. (Suto *et al.*, 1997; reproduced with permission from the *Journal of Chromatography A*).

5.4 Instrumentation

Supercritical fluid extraction is instrumentally demanding, especially when compared with techniques such as Soxhlet extraction or sonication. The first preliminary analytical scale SFE experiments were performed using modified SFC instruments. Only minor modifications are required to adapt an SFC chromatograph for use in SFE. Subsequently, dedicated SFE equipment was introduced by a number of companies. Some of these commercial instruments are still available, others have already been withdrawn from the market. The analytical SFE systems available span a fairly wide range of prices and possibilities. A typical schematic diagram of instrumentation for SFE is presented in Figure 5.7.

5.4.1 Solvent delivery pumps and ovens

Analogous to the situation in SFC, a pump (or a system of pumps) is used to deliver the supercritical fluid. The pump is usually operated in the constant pressure mode. Syringe pumps and reciprocating piston pumps are now the most widely-used fluid delivery systems, although the use of gas compressors or pneumatic amplification pumps has also been described (Pariente *et al.*, 1987). When CO_2 is used as the extraction fluid, the extractant is taken from a cylinder containing this fluid in the

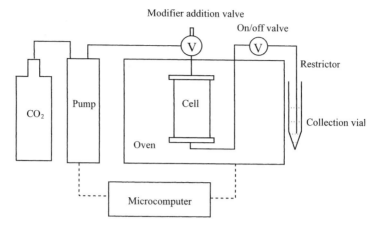

Figure 5.7 Schematic representation of an instrument for SFE.

liquid state under its normal vapour pressure (approximately 55 bar). The fluid is taken directly from the gas cylinder. To facilitate pumping, cooled pump heads can be used or, alternatively, the CO_2 can be taken from a cylinder pressurised by helium, a so-called helium headspace cylinder. The latter option is now increasingly discouraged as the uptake of varying amounts of helium into the CO_2 can affect its dissolution characteristics (Leichter *et al.*, 1996).

The instrumentation schematically depicted in Figure 5.7 shows a single pump SFE system. For extractions that require modified CO_2 as the extraction fluid, more complicated instrumentation may be required, depending on the method that is selected for addition of the modifier. Three methods of preparing modified fluids for SFE have been described in the literature. The simplest is to spike the modifier onto the matrix. A clear advantage of this method is its simplicity, making it attractive for purposes of method development. A disadvantage is that the modifier is rapidly eluted from the cell. Repeated spiking can be an option if the modifier is required throughout the entire extraction experiment. When this method is applied, conditions can be encountered in which phase separation occurs. Fortunately, for most supercritical fluid extractions this does not result in major experimental difficulties. Addition of modifier by spiking is, hence, a rapid method for preliminary scouting experiments.

A second method for SFE with modified fluids is the use of cylinders containing a premixed binary (or ternary fluid). This method is limited to modifiers that have a good solubility in the liquid CO_2 in the cylinder.

For example, the solubility of methanol in liquid CO_2 ($\approx$55 bar) is about 14%. Hence, the maximum concentration of methanol that can be prepared in this way is 14%. For acetonitrile and chloroform, this is only about 2% (Maguire and Denyszyn, 1988). In addition to the high price of binary cylinders, a distinct disadvantage associated with the use of such premixed fluids is the continuous change in the composition of the residual liquid in the cylinder. The most flexible method for the preparation of binary fluids is analogous to high pressure mixing in LC. Here, the effluents from two pumps are combined and mixed prior to introduction into the extraction cell. In contrast to the other methods described above, this method is very flexible with regard to the modifier concentration and even allows the use of composition gradients. A drawback is that two pumps are required.

The second important part of the SFE instrument is the oven, which contains the extraction cell and a preheating coil through which the fluid is pumped prior to introduction of the extractant into the extraction cell. The technical features of the SFE oven are comparable to that of instrumentation for GC, with the exception of the maximum temperature. In SFE, temperatures above 200°C are rarely used. The lower temperature limit is generally slightly above room temperature. High demands are imposed on the temperature stability of the SFE oven as the elution strength of the extractant can be strongly temperature-dependent, especially when extractions are performed close to the critical point of the (mixed) extraction fluid used. Other important parts of the instrument are the extraction cell, the static dynamic valve and the flow-regulation device. The extraction cell and the flow-regulator will be described in more detail in Sections 5.4.2 and 5.4.3. The static/dynamic valve allows selection between the static extraction mode, in which the matrix is 'soaked' with the extraction fluid, and the dynamic flow-through mode, where the extractant percolates through the packed sample bed.

5.4.2 *Extraction cells and cell-packing procedures in SFE*

The extraction cells used in SFE bear a close resemblance to HPLC columns. Indeed, the requirements in terms of pressure resistance and leak-tightness that have to be met by an SFE vessel and an HPLC column are highly comparable. It is therefore not surprising that the extraction cells used in initial analytical scale SFE experiments were empty HPLC columns. Subsequently, vessels specially designed for analytical SFE have been introduced, differing mainly in the way they can be tightened. Special extraction vessels for SFE can be manually tightened and sealing rings and frits can be easily replaced. When using proper sealing procedures, these systems can be leak-tight even for the low-viscosity fluid CO_2 up to very high pressures (500–600 bar). Extraction cells for SFE typically range in size from 150 μl to 50 ml and are constructed from stainless steel or materials of similar inertness (Dean and Kane, 1993). Generally, flow-through, cylindrical cells are used, where the fluid is introduced at one end and leaves the cell at the opposite side.

Orientation of the cell (horizontally versus vertically) and cell dimensions are parameters that can affect the extraction process (Rein et al., 1991; Langenfeld et al., 1992). For example, Furton and Lin (1992) studied the influence of the extraction cell dimensions on the SFE of PAHs from model substrates using cells with different diameter-to-length ratios. For the largest PAHs studied, the SFE efficiencies observed increased by more than a factor of two when the vessel diameter-to-length ratio was varied from 1:20 to 1:1. This is clearly in contrast to expectations. In addition to cell dimensions and orientation, the method of packing the sample matrix into the extraction cell can influence the efficiency and reproducibility of SFE (Richard and Campbell, 1991). Prudent practice is to completely fill the extraction cell, since the influence of packing structure in this case is greatly reduced. Cell orientation is important if the cell is not full, but less important if it is. However, for polymeric samples, problems can occur if the extraction cell is densely packed with a polymer that can swell in the extraction fluid. In such situations, the polymer can be extruded from the cell during extraction or the extraction cell can be blocked by the swollen polymer. When extracting such polymers, the extraction conditions should be carefully selected to avoid excessive swelling and eventual dissolution. Additional safety measures can include partial filling of the cell to a level that is low enough to accommodate any swelling that may occur.

When working with partially filled cells, it is generally advisable to fill the empty part of the extraction cell with a compressible support, such as clean glass wool. This technique of 'multiple layer filling' can also be applied to other matrices and/or for other reasons. For example, placing glass beads (50–100 μm) at either end of the extraction cell can be used to minimise the dead volume of the cell as well as to prevent blockage of the outlet frit of the cell or the restrictor (see Section 5.4.3) (Richard and Campbell, 1991). Wet samples may be mixed with a drying agent or the drying agent can be applied as a separate layer to prevent blockage of the restrictor (King, 1990). Addition of copper to the cell was recommended for the extraction of sediments with high elemental sulphur contents (Bowardt and Johansson, 1994). Silica can be used to selectively remove fats in the extraction of pesticides and PCBs, from e.g. lyophilised fish samples. Finally, Berg and co-workers (1993) reported an on-line reaction and extraction method for fatty acids and glycerides, in which the sample matrix was packed together with different layers of reacting, supporting and dehydrolyzing materials.

In the overview of cells and cell filling techniques presented above, it is assumed that the matrix to be extracted is a solid material. This is the case for the vast majority of samples currently processed by SFE. Although direct SFE of liquid samples, such as fruit juices or even water, has been demonstrated in the literature and special cell designs for the extraction of fluid samples have been described (Hedrick and Taylor, 1989; Ong et al., 1990), better and more reliable techniques, such as solid phase extraction (SPE) or membrane extraction, are available for the routine analysis of such samples. SFE, however, can be used for indirect extraction of samples such as those referred to above, for example after applying the liquid sample to be analysed onto an SPE cartridge or glass beads (Liu and Wehmeyer, 1992; Wright et al., 1987a). Supercritical desorption of the analytes from the SPE cartridge instead of normal liquid desorption offers the following advantages: increased selectivity; reduced usage of organic solvents; and ease of preconcentration.

5.4.3 Flow restrictors

One of the most important parts of an SFE instrument is the restrictor. This part of the instrument controls the mass flow of supercritical fluid through the cell. Eventually, the supercritical fluid is depressurised through the restrictor to ambient pressure conditions. The restrictors used in SFE can be divided into two general categories: fixed restrictors and variable flow devices. Most of the technology for fixed restrictors comes from SFC. Linear fused silica or metal restrictors, pinched restrictors, tapered restrictors, frit restrictors, etc. are simple, low cost devices that are easy to install and operate. Variable flow restrictors are much more complicated. These devices comprise a variable nozzle, where the diameter of the orifice can be adjusted to obtain the desired flow rate. The more sophisticated variable flow restrictors contain a feedback loop, in which the flow rate is monitored continuously and the orifice diameter is varied to keep the flow rate constant at the preselected value.

An important aspect associated with restrictor performance in SFE is plugging. In analytical SFE of samples with high contents of extractable components or of wet samples, partial or complete blocking of the restrictor is frequently observed. During depressurization of the supercritical fluid in the restrictor, the reduction of the density of the extraction fluid results in a sharp decrease in the solubility of the analytes, which in turn may lead to analyte precipitation and eventually plugging of the restrictor. With wet samples, the restrictor can easily be blocked by ice formation at the restrictor outlet due to the Joule-Thomson cooling created by the expanding fluid. Restrictor plugging and the resulting flow rate variations are serious problems in SFE, not only because of the costs and time involved in replacing a plugged restrictor but, even more importantly, because of the fact that extraction yields can be significantly affected. Dedicated instruments for analytical SFE, therefore, contain a flow monitor that indicates how much CO_2 has been passed through the extraction vessel. Experiments in which the volume of extractant used is too low should be rejected.

Evidently, plugging problems are more likely to occur in fixed restrictor systems, although they can also occur occasionally in systems equipped with a variable flow restrictor. An overview of possible solutions for preventing restrictor blockage is given below. Most commercial instruments for SFE incorporate one or more of these options: heating of the restrictor; inserting the restrictor into a suitable solvent in which the components have a good solubility; extracting smaller amounts of the specimen; using restrictors with larger orifices or with different length-to-diameter ratios; and using special materials to trap water and/or other extractable matrix components prior to the restrictor.

5.4.4 Solute collection

The collection of the extracted components from the flow of depressurized extraction fluid is an important step in the SFE procedure. Basically, two distinctly different methods can be distinguished: off-line SFE and on-line SFE. In the latter method, the effluent from the SFE sample preparation step is directly transferred into a chromatographic system resulting in an on-line set-up for sample preparation and analysis. Various on-line systems involving SFE as the sample pretreatment method have been discussed in a book on hyphenated systems for SFC and SFE edited by Jinno (1992). SFE has been coupled on-line to all the major chromatographic techniques including GC, HPLC, SFE, thin layer chromatography (TLC) and size exclusion chromatography (SEC). The present chapter is restricted to off-line sample collection, i.e. collection of the extract in a vial, either directly or after some form of intermediate trapping. For a more in-depth discussion of on-line systems involving SFE, the reader is referred to dedicated literature.

The extract collected from an SFE experiment can be analysed by any appropriate technique, either chromatographic or non-chromatography-based. Multiple analysis is also possible. In off-line SFE, the collection of the extracted solutes is a very important step. Significant losses of the analyte can occur during this step, leading the analyst to believe that the actual extraction efficiency was poor. In recent years, the understanding of the factors influencing off-line solute collection has greatly improved. There are three options for off-line collection: (i) expanding the fluid into an empty container with or without cryogenic cooling; (ii) collection in a (cooled) solvent; and (iii) trapping onto a solid surface or a suitably selected adsorbent (followed by liquid desorption).

The first method was used in the early development of analytical SFE (Wright et al., 1987b). With this method, sample losses can occur during collection due to aerosol formation. A better alternative is method two, in which the supercritical fluid is expanded into an organic solvent. A clear advantage of this method is that the extract is immediately available for further analysis. With this method, efficient collection can sometimes be a problem when high extraction flows are used. This is especially the case when the collection fluid is viscous, for example if a viscous collection solvent or a high-viscosity modifier is used. A sharp increase in the viscosity of the collection fluid can occur as a result of trapping of large amounts of extracted compounds in the solvent. Viscous trapping fluids can be bubbled out of the collection tube. Moreover, an increasing viscosity can also result in slow mass transfer from the gas phase into the solvent. An additional problem of the solvent collection approach is that the volume of the collection solvent can decrease due to evaporation. Not only can this result in incomplete trapping towards the end of the extraction process, but in quantitative analysis partial evaporation complicates calculation of the concentration of the target compound in the matrix. This problem can be solved by adding an internal standard to the collection vial or by gravimetrically determining the volume of collection solvent remaining after the extraction. Cooling the extraction solvent can be an option if highly volatile collection fluids are used.

Although the third method, trapping on an adsorbent, is somewhat more complicated from the experimental point of view, it is generally preferred because it is the most suitable method for automation and, consequently, for routine application of SFE. Other advantages of this method include the possibility of introducing additional selectivity into the sample preparation method by using special sorbents. Moreover, high supercritical fluid flow rates can be used while still maintaining a good collection efficiency. Finally, it has been demonstrated that SFE with solid phase trapping has the potential for simultaneous extraction, clean-up and concentration of the components from different matrices (McNally, 1995; Mulcahey and Taylor, 1992). An important parameter to consider when using adsorbent trapping as the method for isolating the components of interest from the expanded fluid stream, is the nature of the packing material and the temperature of the adsorbent trap. This is especially true when modifiers are used. If excessive amounts of the modifier condense in the trap, the modifier can remove trapped analytes. To avoid condensation of the modifier, the trap temperature must be selected in the proximity of the boiling point of the modifier. The use of trapping temperatures that are too high, however, can result in loss of volatile components.

For many applications, a single collection step will suffice, although tandem combinations of two adsorbent traps or of two different methods might sometimes be necessary (Raymer and Pellizzari, 1987; Vejrosta et al., 1994; Moore and Taylor, 1995; Meyer et al., 1993). Another source of errors in the adsorbent method for component isolation is the volume of liquid delivered by the SFE instrument for desorbing the trap. At an average SFE time of 45 min, the desorption pump has to deliver a small volume of solvent, typically 1–10 ml, once in every 45 min. Start-up problems inherent to most types of pumps result in fairly large errors in the volume delivered. This again might make the use of an internal standard or a gravimetric method for determining the actual solvent volume delivered inevitable.

5.5 Method development

Optimisation is, perhaps, the area of greatest concern in SFE, since it impacts on the accuracy and precision that can be obtained with the technique. Unfortunately, many parameters can affect the extraction process in SFE, among these the most important are temperature, pressure, modifier type and concentration, as well as solute parameters, such as molecular weight, polarity and volatility, and matrix properties, such as particle size, pore structure, water content and adsorptive strength. The multitude of parameters that influence the extraction process and the lack of fundamental knowledge about how these parameters affect the extraction are to blame for the fact that, for a long time, method development in SFE has been mainly empirical. Due to this lack of knowledge, optimisation strategies for SFE have had to be based on trial and error experiments. Although this statement is in part still correct, the significant improvement in the various mechanisms underlying SFE in recent years has greatly reduced the time required for the development of new SFE methods. In the following sections, a number of parameters that play a role in the SFE process will be discussed with the aim of aiding the analyst in the development of SFE methods.

5.5.1 Kinetic models for SFE extraction

An SFE extraction basically involves three subsequent steps. Firstly, the solutes have to diffuse from the core of the particles to the surface. It is obvious that this step is absent if the components are adsorbed on the surface of a particle, which is, for example, the case for components that are extracted from an SPE cartridge or from sandy soil. Next, the components are transferred from the particle surface into the extraction fluid. The key parameter that controls this process is the distribution coefficient of the solute between the matrix and the supercritical fluid. Finally, the components are eluted from the extraction cell by the flow of supercritical extractant. The last two steps of the extraction process are somewhat similar to the processes occurring in SFC. The SFE extraction of a polymeric particle is schematically presented in Figure 5.8. Components diffusing from the core of a particle to the surface are then transferred through a layer of stagnant fluid present around the particle into the flowing fluid phase that carries the solutes out of the cell. The actual extraction rate is determined by the slowest of these three steps. Identification of the rate-determining step is, therefore, an important aspect in method development for SFE.

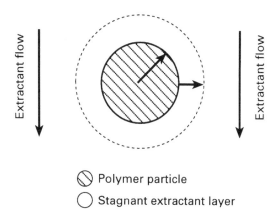

Figure 5.8 The extraction of a low molecular weight material from a polymer particle.

The two possible extremes concerning the rate-limiting step are schematically represented in Figure 5.9.

In Figure 5.9A, a situation is depicted in which the rate-limiting step is diffusion of the solute inside the particles that are extracted. Increasing the diffusion coefficient of the solute in the matrix by, for example, increasing the extraction temperature will increase the extraction rate. If the rate-limiting step is elution of the components out of the extraction

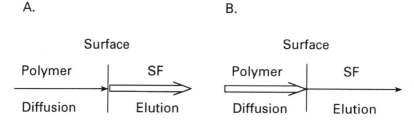

Figure 5.9 Schematic diagram of the two possible extremes concerning the rate-limiting step in SFE. A) The rate-limiting step is diffusion inside the particles. B) The rate-limiting step is elution of the components out of the extraction cell. Abbreviation: SF, supercritical fluid.

cell (Figure 5.9B), enhanced extraction rates can be obtained either by increasing the solvent strength of the extractant (e.g. by increasing the extraction pressure or by the addition of a modifier) or by increasing the flow rate of the supercritical fluid.

Optimisation in SFE has been the subject of a large number of papers in chromatographic literature (Bartle *et al.*, 1989; King *et al.*, 1989; McNally and Wheeler, 1988; Wheeler and McNally, 1989; Furton and Lin, 1993; Shilstone *et al.*, 1990; Giddings *et al.*, 1968; King and Friedrich, 1990; Stahl *et al.*, 1978; Rein *et al.*, 1991; Pawliszyn, 1993). Only a limited number of these publications concern more fundamental studies, in which attempts have been made to increase the basic knowledge of the thermodynamic and kinetic parameters that impact on SFE. Bartle and co-workers (1989) derived a model that allowed the calculation of the extraction kinetics in an SFE system where the rate-determining step is slow mass transfer inside the matrix particles. King (1989) published a mathematical model that enables the calculation of the solubility of organic compounds in supercritical fluids from molecular parameters. Lou and co-workers (1995a) presented a theory that allows the estimation of the SFE extraction recoveries versus time for solutes adsorbed on an SPE packing or a sandy soil. Bartle's model, an adapted hot ball model, yielded an excellent agreement between theory and practical results for samples where the rate-limiting step in the extraction process was diffusion inside the solid particle. A prerequisite for the model to be accurate is that the matrix particles are spheres of a well-defined size and that the initial distribution of the solutes within the spheres is uniform. The extraction of polymer additives or oligomers from polymeric particles is typical of an SFE application where these conditions are met. For such samples, Bartle's models have clearly proved useful. A further refinement of the model incorporated a correction for solubility limitations in the extraction fluid (Bartle *et al.*, 1990).

Despite all attempts to provide better descriptions of the extraction kinetics in SFE, at present they can only be predicted accurately for the two sample extremes, i.e. components homogeneously distributed throughout spherical solid particles of identical size or components adsorbed onto the outside of a well-defined matrix. For all other samples, SFE conditions have to be optimised experimentally. Further general guidelines for this optimisation will be presented in the following sections.

5.5.2 Sample morphology

In extraction processes, the actual extraction behaviour depends on the solutes under investigation as well as on the properties of the matrix. The ideal sample for any type of extraction method is a dry specimen, reproducibly taken from the entity to be analysed, consisting of small particles, homogeneous in size and chemical composition. In this ideal sample two extremes can be distinguished. At one extreme, the components of interest are homogeneously distributed through the matrix particles. Polymer samples ground to a well-defined size can approach this extreme. At the other extreme, the components are adsorbed onto the outside of the particle with homogeneous adsorption energies. To visualise this type of sample, one could think of an adsorbent onto which the components of interest are adsorbed. Most real-life samples are significantly different from the ideal sample defined here. Environmental samples, for example, are generally wet, contain particles of different sizes and chemical composition, and the target solutes can be present partly inside this wide range of particles and partly on adsorptive sites of widely varying energy on the outside of the particle. The ease with which the components can be removed from each of these greatly differing and nonhomogeneous environments can be strongly dependent on the concentrations of the components in the various compartments, the moisture content, etc. Sample morphology is therefore an important parameter in developing SFE methods.

5.5.3 Sample size

The sample size is the first parameter that should be given careful consideration before performing an SFE extraction. First, the sample subjected to extraction should be large enough to ensure sample homogeneity and to obtain sufficient sensitivity. In this respect, SFE appears to be similar to other methods of sample preparation. SFE has, however, unique experimental difficulties. If larger amounts of sample are processed, this can easily result in instrumental problems. Large amounts of sample require larger amounts of supercritical fluid for quantitative extraction, may easily block the restrictor and can make trapping of the extracted analytes more difficult, especially for volatile components. Moreover, large extraction cells can present a serious safety hazard. Taking all these together means that, in general, a small sample size is preferred so that the requirements for sample homogeneity and sensitivity are satisfied.

5.5.4 Selection of the supercritical fluid

The properties of supercritical fluids have been addressed in detail in Section 5.2. As reported there, CO_2 is the first choice as the extraction fluid for SFE. Many of the stated advantages of SFE, such as environmental- and operator-friendliness and low cost, are in fact directly related to the use of CO_2 as the extraction fluid. In general terms, CO_2 is an excellent extraction medium for nonpolar to moderately polar species. It has been successfully applied to extract a wide variety of compounds, such as alkanes, PAHs, PCBs, fats, esters and pesticides, from various matrices. However, CO_2 does not have sufficient solvent strength to dissolve more polar compounds. Moreover, in the SFE extraction of real-life samples, less than quantitative extractions were frequently reported, even for only moderately polar compounds, when using CO_2 as the extraction fluid. In such samples, the analytes are adsorbed on active sites of the sample matrix and CO_2 is a poor fluid to overcome the matrix/solute interactions. To achieve quantitative recoveries, the ability of the supercritical fluid to overcome these interactions is often more important than a high solubility.

Due to the lack of acceptable, more polar, single-component fluids for SFE, modifiers are widely used when extraction using pure CO_2 fails. Many different modifiers have been used in SFE. Unfortunately, little information is available to aid in the selection of modifiers and their concentrations. The solubilities of the analytes in the modified supercritical fluids, and the interactions between the modified supercritical fluid, the matrix and the target analytes are poorly understood. In selecting modifiers, properties both of the target solutes and of the matrix should be considered. Until the action of modifiers is better understood, the choice of the best modifier for the extraction of complex samples is based on empirical experience. For a tentative survey of modifiers, some preliminary guidelines can be utilized (Lou, 1997):

(i) In the extraction of components which are highly soluble in supercritical CO_2 from homogeneous matrices (where the components are only weakly adsorbed on the matrix surface and the

matrix does not contain active sites), modifiers are virtually not needed (for example, in most spiking/recovery studies).
(ii) The modifier should not interfere with the subsequent analysis.
(iii) In the extraction of inhomogeneous samples (containing active sites, such as in most environmental samples), polar or reactive modifiers can be necessary to deactivate active sites on the surface of the sample matrix, even for the extraction of compounds which are highly soluble in supercritical CO_2. In this case, a small amount of a suitable modifier may result in a significantly improved extraction yield.
(iv) For polymeric samples, where modifiers are used mainly to increase diffusion of analytes from the core of the polymer to its surface, the modifier should be a good swelling agent for the polymer. Quite frequently, large amounts of modifier are necessary and continuous addition of modifier or repeated spiking of the modifier may be of benefit.
(v) For components which are not soluble in CO_2, the modifier should be a good solvent for the target analytes, or a reactive modifier (derivatizing reagent) should be used to transfer the compounds into extractable, nonpolar derivatives.
(vi) For ionic compounds, ion-pairing or complexation modifiers should be used.

Optimization of SFE methods using modified fluids frequently requires time-consuming experimentation in which modifiers with different polarities and concentrations are evaluated at various pressures and temperatures (Hawthorne, 1990). In this process, the six guidelines presented above can prove valuable.

5.5.5 Pressure

The first and most obvious requirement of the SFE conditions selected is the ability to dissolve the target analytes. The solvent strength of a supercritical fluid can easily be controlled by changing the pressure and/or, to a lesser extent, the temperature (King et al., 1989). Pressure is one of the main parameters that influences extraction recovery and selectivity in SFE. An increase in pressure at a constant temperature results in an increase in solvent strength, which means a better solubility of the solutes in the supercritical fluid. In addition, the higher the extraction pressure, the smaller the volume of supercritical fluid needed for extraction (McNally and Wheeler, 1988). In method development, a rapid experiment is an initial experiment with an extraction pressure equal to the maximum pressure of the SFE system. Such an experiment gives valuable information on the feasibility of the extraction. Under these conditions, a high recovery should be obtained in a short extraction time. If under these conditions the target solutes cannot be extracted, it is likely that modified extraction fluids might be required. High pressures, however, cannot always be recommended because of the limited selectivity of SFE at high pressures. Selective extraction can be achieved only by a proper choice of extraction conditions. For example, alkanes can be extracted from urban air particulates with CO_2 at 75 bar (45°C), whereas the PAHs remain unextracted until the pressure is raised to 300 bar. By sequentially extracting the air particulates at these two pressures, 85–90% selectivity can be achieved (Hawthorne and Miller, 1986). Successful selective extractions have also been reported for extracting analytes from bulk matrix materials that are also soluble in the supercritical fluid under stronger conditions (Hierro and Santa-Maria, 1992).

5.5.6 Temperature

At a constant pressure, the density of a supercritical fluid decreases when the temperature is increased. On the contrary, temperature can also affect the volatility and diffusivity of the solutes, the flexibility of the matrix, and the affinity of the solutes for active sites on the matrix. Hence, the effect of a temperature elevation on SFE recovery and selectivity is difficult to predict (Krunic et al., 1981). On the other hand, for a volatile solute there is competition between its solubility and volatility, and higher recoveries can be obtained at higher temperatures (Wheeler and McNally, 1989). If the rate-limiting parameter in SFE is diffusion of the components or desorption of the analytes from active sites of the matrix, increasing temperature will result in an improved extraction recovery because increased temperature will facilitate solute diffusion and/or reduce the interaction between the solute and the matrix. For example, in the extraction of polymer additives from polymeric materials, an increase in temperature normally gives a faster extraction (Lou et al., 1995b; Cotton et al., 1993a). Langenfeld and co-workers (1993) reported that increased extraction efficiencies of PCBs from sediments and of PAHs from air particulate materials were obtained by increasing the extraction temperature from 50 to 200°C at 350 bar. According to Langenfeld and co-workers these results indicate that the kinetics of the partitioning process are improved at higher temperatures (Langenfeld et al., 1993). However, the improved extraction efficiency observed could also be due to a reduced interaction between the solutes and active sites at higher temperatures (Janssen et al., 1989).

5.5.7 Flow rate and extraction time

The way in which a change in extraction flow rate affects the extraction process is key to understanding the extraction kinetics in SFE. Experiments at different flow rates can yield valuable information about whether the major limitation to achieving rapid extraction is of a thermodynamic nature (i.e. the distribution of the analytes between the supercritical fluid and the sample matrix at equilibrium) or is related to kinetics (i.e. the time required to approach that equilibrium). For samples that show a significant increase in extraction rate when the fluid flow rate is increased, the kinetics of extraction are apparently fast. Solubility appears to be the limiting factor, and the extraction can be improved by increasing the extraction pressure or by exposing the sample to a larger volume of fluid. In contrast, if there is no large effect of the fluid flow rate on the extraction rate, it appears that mass transfer is slow, and this slow mass transfer limits the overall extraction rate (Hawthorne *et al.*, 1993). In this case, an increase in the extractant flow rate does not affect the extraction rate. In SFE with CO_2 as the extraction fluid, typical supercritical fluid flow rates range 0.1–1 ml/min.

The optimal extraction time depends on the experimental temperature and pressure, the modifier type and concentration, and the flow rate of the fluid through the extraction cell. For extractions where modifiers are added directly to the sample matrix, a period of static extraction is often found to be necessary to ensure that the modifier makes good contact with the solutes and the matrix. The static times used generally range 5–30 min, depending on the properties of the solutes, the modifier and the matrix. For unknown samples, the extraction time can best be found by experimentally conducting successive extractions to determine the completeness of extraction. The use of a nondestructive detector in tandem with SFE can also aid the analyst in determining the extent of extraction. Knowledge of the sample matrix and the solubility of the components in the supercritical fluid can be of assistance in choosing the proper extraction time, since the extraction recovery is a function of the ratio of a solute distributed between the supercritical fluid and the matrix. For example, extraction times of more than one hour are always needed in the extraction of polymeric materials, while complete extraction could be obtained within one hour for most environmental samples under properly selected conditions.

5.5.8 Collection

The methods of solute collection in SFE have already been discussed in detail in Section 5.4.4. The collection of the extracted components is a very important step in the SFE procedure. Accurate evaluation of the efficiency of extraction can only be made if collection is quantitative. Hawthorne and co-workers (1993) suggested that the collection efficiency should be the first parameter to be studied in developing an SFE method. Because of the large number of experimental variables that can affect collection efficiencies, the quantitative abilities of the collection device must be determined using appropriate spiking/recovery studies. In these studies, the conditions to be used for the real-life samples should be used to extract the analytes of interest spiked at known concentrations onto a relatively inert matrix, such as clean sand. The matrix should retain the spiked analytes until the SFE is begun, but should easily release the spiked analytes during the extraction, since the goal is to evaluate only the method of collection. Fortunately, even with relatively simple methods of collection, such as bubbling the depressurized extraction fluid through a few millilitres of solvent, quantitative collection of analytes as volatile as n-octane and phenol are relative easy to achieve (Langenfeld *et al.*, 1992; Burford *et al.*, 1992; Porter *et al.*, 1992). When more volatile analytes are of interest, or if very high supercritical fluid flow rates are used, the use of sorbent-trapping should be considered. Sorbent-trapping, as has been stated previously, is also easier to automate.

5.5.9 *Optimizing and validating SFE methods*

The development of an SFE method is a time-consuming process, because it is affected by many parameters. The mechanical and physicochemical aspects of SFE that have been discussed above can only provide some general guidelines in developing quantitative SFE methods. Hawthorne and co-workers (1993) proposed a sequential process for the development of SFE methods for environmental samples. The value of this process has been clearly demonstrated for SFE method development in our laboratory.

In the selection of initial SFE extraction conditions, the properties of the target analytes and matrix and any literature reports on successful SFE methods for similar samples should be considered. The initial selections will rarely be satisfactory because extraction efficiencies in SFE are highly matrix-dependent. The following are some suggestions for selecting the initial conditions in SFE experiments:

(i) Pure CO_2 can extract nonpolar to moderately polar components that can be analyzed with conventional capillary GC.
(ii) For polar or highly nonvolatile components, the addition of modifiers may be necessary to obtain quantitative extraction. For

example, ion-pairing reagents are needed for extracting ionic compounds.
(iii) The initial supercritical fluid flow rate and extraction pressure should be 0.5–2 ml/min of the supercritical fluid at the maximum pressure of the SFE system.
(iv) For thermally labile compounds, low extraction temperatures are favourable.
(v) In general, supercritical fluids are more effective extraction agents when the extraction is performed at a temperature above the melting points of the target analytes.
(vi) In the extraction of polymeric materials, the extraction temperature should be selected well above the glass-transition temperature of the fluid-swollen polymer, in order to increase diffusion of solutes in the polymeric material (unfortunately, the glass-transition temperature of a polymer under supercritical conditions is rarely known).
(vii) For samples with high concentrations of water and/or other extractable matrix components, the restrictor should be heated or some type of drying agent should be added to the sample to avoid plugging of the restrictor during extraction.
(viii) For environmental samples, extraction times of 30–60 min are useful, since longer extraction times do not generally yield substantially higher recoveries. Extraction times of at least one hour are normally needed for polymeric samples.

After the initial selection of extraction conditions, the collection efficiency should first be tested and optimized. Only when the trapping efficiency is quantitative can the extraction efficiency and selectivity be further evaluated. The collection of the extracted components in SFE was discussed in more detail in Section 5.4.4. Since samples with known concentrations are only scarcely available, the validation of an extraction method is generally based on one of the following three approaches, each depending on assumptions that may or may not be valid (Hawthorne et al., 1993): (i) determination of the recovery of known concentrations of spiked compounds from the sample (or similar) matrix; (ii) comparison of the recoveries of native analytes with those achieved using conventionally-accepted extraction methods; and (iii) performance of multiple sequential extractions of the same sample under different extraction conditions.

Perhaps the least reliable technique for validating the efficiency of SFE methods is the use of spiked samples (Burford et al., 1993). It should also be noted that conventional extraction methods may not yield quantitative recoveries of the native analytes, and therefore a highly efficient SFE method may yield higher recoveries than conventional methods. The use of multiple extractions with different conditions could be a very useful way to validate the efficiency of SFE. This could include extracting the residues from an SFE by Soxhlet in an appropriate liquid solvent, by extracting the residue under more stringent SFE conditions, or by a combination of these two approaches (Hawthorne *et al.*, 1993; Burford *et al.*, 1993).

Under the conditions initially selected, SFE extraction may not be quantitative or sufficiently selective, and the optimization step may have to be repeated several times. Experimental design approaches have been used to this end to reduce the time and the number of experiments needed for the optimization of SFE (Adasoglu *et al.*, 1994; Barnabas *et al.*, 1995). An optimization strategy for the extraction of polymeric samples was recently proposed (Figure 5.10) (Lou *et al.*, 1996). Although originally developed for the extraction of polymeric samples, this approach can be applied to other samples as well.

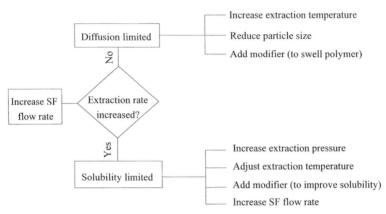

Figure 5.10 Optimization strategies for supercritical fluid extraction (SFE) of polymeric samples. Abbreviation: SF, supercritical fluid. (Lou *et al.*, 1996).

A 'good' extraction method, in addition to quantitative recoveries, gives a high degree of selectivity. It is quite possible that the demands for quantitative and selective extraction may be very difficult to meet simultaneously. Quantitative extraction with recoveries being independent of the matrix requires a strong extraction fluid. Selective extraction, on the contrary, can only be obtained by using fairly mild extraction conditions where only the components of interest are dissolved from the matrix. For certain samples, a compromise will have to be sought. Here the clear advantage of SFE is that it offers the operator the possibility for fine-tuning selectivity, a technique not available, or much more complicated, in many of the other extraction methods.

If the required selectivity cannot be introduced in the extraction step alone, a suitable sorbent can be packed into the extraction cell and the extraction process can be repeated under properly selected conditions. This additional selectivity was demonstrated by Sandra and co-workers (1994) in the analysis of organochlorine pesticides in tobacco leaves. The chromatogram could be greatly simplified and the components of interest easily quantified by adding silica to the extraction cell to retain medium polar and polar solutes.

5.5.10 Using SFC retention data for SFE method development

Although at first glance very different, SFE and SFC are techniques that have many aspects in common. SFC can therefore be used to derive data relevant to method development in analytical SFE (Smith *et al.*, 1987). Useful data for SFE method development that can be assessed by SFC include: diffusion coefficients (in the extractant phase as well as in the matrix); distribution constants; solubility data; virial coefficients, etc. McNally and Wheeler (McNally and Wheeler, 1988; Wheeler and McNally, 1989) attempted to make a direct correlation between SFC retention data for diuron and linuron measured on an octadecyl material with the extraction behaviour of these components from soil in SFE. Similar work was carried out by Furnton and Lin (1993). The information obtained from this work was clearly useful, though only in a qualitative sense. Accurate, quantitative prediction of SFE kinetics from SFC retention is only possible for certain types of samples, in particular those in which the components are present adsorbed onto the outside of a homogeneous particle (Lou *et al.*, 1995a). For the vast majority of real-life SFE samples, however, this prediction is not possible as the extraction kinetics are strongly affected, or even determined, by diffusion inside the matrix particle and/or the concentration of the target solute in the sample. Nevertheless, SFC retention data can still be useful. If a solute requires extreme elution conditions in SFC, extraction in SFE will certainly be difficult. Although almost trivial at first sight, this simple guideline has often been overlooked in the past, resulting in a large series of SFE experiments that could not but fail.

5.6 General areas of application

SFE has been applied to a wide range of samples from numerous areas of application. Three important areas of application of analytical SFE are: environmental analysis, polymer characterisation and food analysis.

Other important areas where SFE has been applied successfully include: pharmaceutical analysis, natural product research and fuel characterisation and classification. Figure 5.11 shows the distribution of SFE methods among the major fields of analysis (Valcarcel and Tena, 1997). From this figure, it is clear that environmental analysis and food characterisation are the most important areas of application. The third largest field of application is industrial analysis. This includes mainly, but not exclusively, polymer extractions. The use of SFE in a number of these areas will be addressed in the following sections. The discussions will be fairly wide-ranging and will focus mainly on the general background of SFE in the particular application area under study, as well as on advantages of the SFE-based methods for sample preparation over other methods.

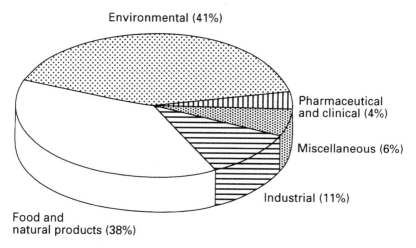

Figure 5.11 Distribution of SFE applications over the various fields of application. (The data presented in this figure were taken from the work by Valcarcel and Tena, 1977).

5.6.1 SFE in environmental extractions

Environmental analysis is without doubt one of the most important and challenging areas of application of chromatography. The components of interest are generally present at extremely low concentrations in a very complex matrix. The range of samples that have to be analysed varies from clean air to highly contaminated industrial waste water and aged sediment. Within one matrix type, the water concentration, organic matter content and concentrations of the target compounds can vary dramatically, making every sample basically a new sample. Moreover, the number of samples that have to be analysed is very large and cost is an important issue.

Looking to these particular problems in the analysis of environmental samples, a number of important advantages of SFE immediately become apparent. Evidently, selectivity is a key issue in environmental analysis and it is advantageous to introduce selectivity as early as possible in the analytical procedure, i.e. during extraction. SFE offers the possibility of introducing selectivity by careful selection of pressure, temperature and modifier type and concentration. In addition to selectivity, SFE also offers increased extraction rates owing to the better mass transfer characteristics of supercritical fluids. Preconcentration, another important and often difficult step in all analytical methods for environmental trace analysis, is not generally required in SFE. Depending on the sample collection procedure applied, little or no organic solvent is used. As there is no need to incorporate an evaporative preconcentration step into the procedure, SFE is relatively fast. A final advantage of SFE in environmental applications is the ease with which this technique can be coupled on-line with the various chromatographic analytical techniques. This can be of benefit when large series of samples have to be processed cost-effectively.

Although SFE is a technique for preparation of solid samples, the other environmental matrices, air and water, are also amenable to SFE.

Air analysis
Two techniques are widely used for the analysis of air samples, i.e. collection of the components of interest in a liquid through the use of an impinger-type device, and trapping of the components onto a suitable sorbent followed by either thermal- or liquid-desorption. Each of the three resulting methods has its own advantages and disadvantages. Liquid-trapping techniques and solvent desorption of adsorbents are difficult to automate and suffer from sensitivity problems. In the more widely-used method of adsorption/thermal desorption, thermal degradation can occur in the case of thermally labile solutes. If liquid desorption of the sorbent is applied, preconcentration of the desorption fluid is often necessary, which can be a tedious procedure owing to the high volatility of the solutes involved. Moreover, liquid desorption is difficult to combine on-line to e.g. GC for further analysis.

SFE can be applied to overcome some of these problems. Using supercritical desorption, low temperatures can be applied and preconcentration is less cumbersome. Moreover, on-line combination with GC is easier, resulting in an automated system and also reducing the risk of sample degradation or losses. Hawthorne and co-workers (1989b) used a polyurethane foam to trap hardwood smoke, phenols from chimney smoke, followed by SFE. Various authors have applied SFE to desorb PAHs present in air (or on airborne particles) from (fibre) filters (Hawthorne et al., 1989a). Swami and co-workers (1997) could detect the termiticites, chlorpyrifos and chlordane, at extremely low levels (0.1 ng/ m^3) in air by sampling large volumes of up to 20 m^3 through a combined trap containing Florisil and foam plugs followed by SFE with off-line analysis.

Liquids
Basically, two approaches exist for the SFE extraction of aqueous samples. The direct 'liquid-liquid' extraction of water is seriously complicated by technological problems, the most critical step being the separation and collection of the two immiscible phases. As a consequence, only a limited number of studies have addressed this direct method for SFE of water samples; these papers have recently been reviewed (Janda et al., 1997). A cell design for SFE of liquid samples was developed by Hedrick and Taylor (1989). In this design of the extraction vessel, CO_2 is bubbled through the water sample. The CO_2 is introduced at the bottom of the cell and leaves the system via an exit at the top; phase separation is obtained through gravity. Thiebaut and co-workers (1989) used a phase separator previously developed for automated continuous liquid-liquid extraction to effect separation between a flowing water phase and the CO_2 phase. The system was successfully applied to the analysis of phenols in surface water.

Most of the applications of SFE to the extraction of liquid samples are performed using an indirect method. Here, the liquid sample is first adsorbed on a solid material before being extracted. The mechanisms involved in the adsorption step can range from immobilisation of the water due to soaking (Hawthorne et al., 1990) to SPE, in which the water is actually removed from the cell prior to starting the extraction and only the components of interest are trapped onto the adsorbent (Liu and Wehmeyer, 1992).

Solid samples
The vast majority of environmental SFE applications concern extractions of solid samples, such as soil, sediment, sludges and air particulates, as well as biotic tissue. A very large number of studies involving numerous classes of solutes have been published in recent years. Solute classes studied include: pesticides and herbicides, solvents, explosives, (nitro)-PAHs, (chlorinated) phenols, oil, etc. The most important advantages of SFE over conventional methods for the extraction of environmental samples, such as Soxhlet and sonication, are: the speed of SFE; the ease of preconcentration of the extract; and the possibility of introducing

selectivity into the extraction process. This third advantage is the most important, as it allows the analyst to reduce the complexity of the sample prior to introduction of the extract in the chromatographic system.

One of the most important disadvantages of SFE is that SFE conditions that successfully extract a specific pollutant from one environmental sample may not yield quantitative recovery from a different matrix. This can be attributed to two parameters: the heterogeneity of real-life environmental samples and the water content. The nature of the matrix has a strong influence on the distribution of the analyte between the supercritical fluid and the sorptive sites of the sample matrix. Water has similar effects and, in addition, can also affect the accessibility of sample pores for the extraction fluid. An important characteristic of the matrix, with distinct consequences for the extraction behaviour of an environmental soil sample, is the organic matter content. For example, the recovery of explosives from soil samples was found to decrease significantly when the organic matter content of the soil increased (Engelhardt et al., 1991).

The effect of moisture on the extraction yields in SFE has received a great deal of attention in the literature. The results of these studies once more indicate that different samples can show a distinctly different behaviour. For example, Pratte and co-workers (1996) studied the influence of water on the SFE of soil. Small amounts of water ($<5\%$) were found to increase the extraction rate of naphthalene from the soil. Water contents of 5–10% had no further effect; however, at water concentrations exceeding 20% extraction yields decreased dramatically. Ashraf-Khorassani and co-workers (1995) studied the influence of moisture on the extraction of PAHs and phenols from soil. The results of their work showed that the presence of water in soil ($>10\%$) increased the extraction efficiency of the higher molecular weight PAHs. In the extraction of phenols, a significantly faster extraction was obtained for soil containing 5% water relative to dry soil or soil containing 1% water.

5.6.2 Applications of SFE in food analysis

Supercritical fluid extraction (preparative as well as analytical scale) is currently finding widespread acceptance in the analysis of foodstuffs, agriculturally-derived materials and natural products. The possibilities and limitations of the technique in this area have recently been the subject of a number of excellent review articles. Lehotay (1997) reviewed the use of SFE for the extraction of pesticides from food, and Valcarcel and Tena (1997) reviewed the state of the art of SFE methods for food extraction, including vegetable and plant samples as well as seeds and animal feeds.

Finally, Eller and King (1996) reviewed the use of SFE for the determination of fat content in foods. SFE has been applied for a very wide range of analytes from an even wider range of food-related samples. A nonexhaustive overview of matrices and target solutes is presented in Table 5.3.

Table 5.3 Foodstuff and related samples studied by supercritical fluid extraction (SFE)

Matrices
Liver, milk, ham, sausage, potato chips, cereals, egg yolk, honey, cheese products, tissue, kidney, fat, beef tallow, (dehydrated) beef, muscle, (lyophilised) fish, seafood, wheat, corn, animal feed, peanut butter, rice, grain, cotton seeds, wheat cake, cornstarch, oilseeds, soybean, vegetables, fruits, potato, tomato, onions, radish, strawberry, mushroom, algae, hops, ginger, spices, lemon peel, etc.

Analytes
Cholesterol, fatty acids, (chlorinated) pesticides, PCBs, PAHs, androsterone, veterinary drugs and residues, anabolic steroids, nitrosamines, total fat, clenbuterol, sulphonamides, fluvalinate, n-alkanes, organotins, methylmercury, aflatoxins, strychnine, organophosphorus pesticides, tocopherols, emulsifiers, triglycerides, flavours, β-carotene, essential oils, waxes, bitter acids, volatiles, pesticide residues, etc.

Abbreviations: PCB, polychlorinated biphenyl; PAH, polycyclic aromatic hydrocarbon.

Food analysis shows a significant similarity with the analysis of environmental samples. In both cases, the analyst often has to determine trace levels of compounds in a very complex matrix containing large amounts of co-extractable material, such as humic acids in the case of environmental samples or fat in the case of food samples. The selectivity of the extraction is therefore an important issue in these two areas of application of SFE.

The co-extraction of fat is an important problem in the SFE of toxic compounds, such as drugs and drug metabolites, pesticides, PCBs, etc., from food samples. Co-solvent addition, high pressures, high modifier concentrations and long extraction times are often mandatory for the extraction of these analytes because of their polar nature (with the exception of the PCBs and the older organochlorine pesticides) and because of the chemical binding of the target solutes to high molecular weight proteins and carbohydrates. Under these conditions, co-extraction of fat can cause serious problems. In general, the working conditions used must be a compromise between efficiency and selectivity, in order to reduce the fat content of the extract. Large amounts of fat are extracted at 80°C and high densities. Fat extraction is dramatically reduced at densities below 0.6 g/ml (Hedrick, 1992). Even under these compromise conditions, an additional clean-up step is often required. Addition of a fat sorbent, such as silica, to the extraction cell can be an elegant way to effect this (Hale and Gaylor, 1995).

In addition to being an unwanted co-extractant, fat can also be the target solute in the SFE of foodstuffs. Conventional methods for determining the fat content, such as ether extraction, can be replaced with SFE methods, since supercritical CO_2 is suitable for dissolving lipids. The speed, the environment- and user-friendly nature of CO_2 and the low costs make the method ideally suited for this application. Quantitative lipid extraction from a variety of meat and cheese products, including meat, sausages, ham, black pudding and various other food products with fat contents ranging 2–50%, and their gravimetric determination can be accomplished in approximately one-tenth of the time taken by classical solvent extraction. Typical SFE recoveries (versus liquid extraction) were greater than 95%, with relative standard deviations (RSDs) of 0.1–1% (King et al., 1989; Lembke and Engelhardt, 1993). Fractionation of the lipids is possible by changing the pressure and using a modifier. For example, phospholipids are not extracted in pure CO_2, but they are extracted in 5%-modified CO_2 (Temelli, 1992).

5.6.3 Extractions of polymeric materials

Polymeric materials consist of the polymer itself in addition to a series of smaller molecules. These include residual solvents and monomers, as well as additives, such as plasticisers, antioxidants and light stabilisers, added to improve the polymer properties or prolong the lifetime of the material. In addition to this, there may be processing aids and inadvertent contaminants present. The accurate quantification of the levels of these materials is important for the manufacturers and the regulation institutions to ensure the product meets its specification. Food products, for example, are subject to strict regulations with regard to the maximum concentrations of low molecular weight material present in the plastic. The presence of large amounts of additives or other low molecular weight material in food-packaging material is a potential source of contamination of the food.

Basically, the techniques for sample preparation for the determination of low molecular weight solutes in polymers can be divided into two categories: dissolution of the polymer (e.g. Crompton, 1984) (often followed by reprecipitation of the polymer); and (liquid/solid) extraction methods. SFE clearly belongs to the latter class. The dissolution/reprecipitation method, if feasible, should generally be the sample pretreatment method of choice in polymer analysis. It is extremely fast, cheap and uses only minor quantities of solvent. Possible complications, however, could be: inclusion of the target solutes in the polymer network during reprecipitation; or the presence of large amounts of nonelutable compounds (e.g. heavy waxes) in the sample after reprecipitation (Spell and Eddy, 1960).

SFE has been applied to the extraction of a very wide range of low molecular weight compounds from an equally wide range of polymers, including: standard polymers, such as the Nylons (Lou et al., 1996; Jordan et al., 1997; Venema et al., 1993); poly(butylene or ethylene terephthalate) (Kueppers, 1992; Cotton et al., 1993a; Schmidt et al., 1989); poly(vinylchloride) (Hunt and Dowle, 1991; Marin et al., 1996); polystyrene (Jordan et al., 1997; Burgess and Jackson, 1992); polyethylene (Ashraf-Khorassani and Levy, 1990); and less common materials, such as biodegradable lactide co-glycoside polymers (Braybrook and Mackay, 1992). Important parameters in the SFE of low molecular weight material from polymers are the particle size, the swelling behaviour and, evidently, the extractant pressure and temperature.

As discussed in Section 5.5.1, the rate-determining step in the extraction of polymeric particles is generally diffusion inside the polymer. Diffusion in polymers is a slow process. Method development for the SFE extraction of polymers should, therefore, be directed to finding conditions under which diffusion in the polymer particle is enhanced. This can be achieved via temperature variation and swelling. Polymers can swell significantly when in contact with CO_2, containing or not containing a modifier. This swelling can result in a significant enhancement of the extraction rates in SFE. Data by Cotton and coworkers (1993b) indicate that the diffusion coefficients in swollen polymers can be up to two orders of magnitude larger than in the nonswollen material.

Together with the reduction of the particle size, physically swelling the material is a very effective method of improving the extraction rate in SFE of polymeric samples. Lou and co-workers (1996) concluded that in the selection of the extraction conditions for SFE of polymers, conditions should be selected that result in maximum swelling of the material. The amount of CO_2 and modifier that will partition into a polymer particle depends on pressure, temperature and the properties of the polymer. The most significant swelling is exhibited by many amorphous polymers. Crystalline polymers are less amenable to swelling by CO_2, probably because of the low solubility of CO_2 in the crystalline regions of polymers (Michaels and Bixler, 1961). Modifiers generally accelerate extraction when they interact with the polymer more than CO_2. The modifier can cause greater swelling than the CO_2 alone. The best results are obtained if the nature of the modifier matches that of the polymer. The authors (Lou et al., 1996) investigated the influence of modifier polarity on the extraction behaviour of caprolactam from Nylon-6, a polar polymer, and of the cyclic trimer from polybutyleneterephthalate (PBT). The results of this experimental study are summarised in Table 5.4.

Table 5.4 Modifier effects in the extraction of caprolactam from Nylon-6 and trimer from polybutyleneterephthalate (PBT)

	Caprolactum from Nylon-6	Trimer from PBT
CO_2	100	100
CO_2 + methanol	310	214
CO_2 + chloroform	182	510
CO_2 + benzene	122	281
CO_2 + hexane	122	123

Extraction conditions: pressure 300 bar; time 20 min static, 30 min dynamic; modifiers were spiked onto the polymer prior to commencing extraction (0.5 ml); temperature 90°C; collection solvent 5 ml dichloromethane; restrictor 50 μm × 70 cm fused silica capillary. The extraction yields listed are relative to those observed for pure CO_2.

The data in Table 5.4 clearly indicate that methanol is the most effective modifier for the more polar Nylon-6 polymer, whereas chloroform is more effective for the extraction of the cyclic trimer from PBT. By performing the same experiment at a range of temperatures, it could be shown that the observations described in Table 5.4 were due to swelling of the polymer rather than to improved solubility of the target analyte in the mixed mobile phase. The modifier effect was most pronounced in conditions of lower temperature, because diffusion in the nonswollen polymer is slowest at low temperatures.

The effect of temperature on the extraction of oligomers from polyethylene terephthalate (PET) was studied by Kueppers (1992). Extraction at a constant CO_2 density of 0.5 g/ml showed an increase in the rate of the extraction of the trimer in the temperature range 40–160°C. In this application, increasing the density at constant temperature resulted in increased extraction rates, indicating that here also solubility is limiting the extraction. The importance of fine-tuning pressure as a method of obtaining selective extractions was pointed out by Engelhard and co-workers (1991). When extracting polyethylene at 152 bar and 45°C, erucamide could be efficiently isolated from the polymer. By increasing the extraction pressure to 203 bar, an Irganox-type antioxidant could also be extracted.

SFE has attracted considerable attention in the polymer industry. This is due, to a large extent, to its much shorter extraction time in comparison with Soxhlet extraction. Additional advantages are the low cost and environment-friendly nature of the extraction solvent in addition to the easy automation of the technique. Easy extract preconcentration and the possibility of introducing selectivity through the choice of extraction parameters, generally speaking, are less important issues in this area of application. In method development for SFE of polymeric materials, conditions should be selected that enhance diffusion inside the particles, as this is normally the rate-determining step. Care should be taken not to exceed the softening or melting point of the polymer. Once optimised, large series of samples can be extracted in a fully-automated way. As variation between the samples is generally small, optimised conditions will give rise to reproducible yields for all samples from a series. Important disadvantages of SFE over Soxhlet extraction include the much higher price of the equipment and the less robust nature of the technique.

5.7 Future directions and developments

SFE has been promoted by researchers and manufacturers as the ideal technique for sample preparation for chromatography. It was said that SFE: provided the selectivity required for even the most complex samples; used no organic solvents; was fast; required no tedious extract preconcentration; was applicable to a very wide range of samples; was simple to automate; was easy to couple on-line to the chromatographic analysis system, etc. It is now clear that all this is far too optimistic. Some of the statements quoted above simply cannot be true for fundamental reasons. For example, only techniques that apply mild extraction conditions or use some type of very specific chemistry can be selective. Mild extraction techniques, however, are restricted in their range of application and are likely to show matrix-dependent extraction yields. Selective SFE can be performed by applying mild extraction conditions, but under these conditions the technique is likely to suffer from a strong matrix-dependency of the extraction yields, just as any other mild extraction method would. SFE is somewhat faster than Soxhlet, but with Soxhlet extraction 20 or more samples can be extracted in parallel, which is clearly permitted by the price of Soxhlet equipment. Moreover, the SFE extraction of certain samples can be fast, but method development in SFE is a complicated process. Hence, a considerable net gain in time is only achieved for samples that have to be analysed in large series. From all this, it can be concluded that SFE indeed holds a number of potential advantages; however, expectations should be realistic. It is almost fundamentally impossible that all the above-mentioned advantages occur simultaneously.

SFE was clearly not a mature technique when it was introduced. For new techniques to become accepted, a good understanding of the basic principles should be available together with reliable instrumentation to make the new method work. Moreover, expectations should be realistic. When SFE was introduced, the understanding of the mechanisms underlying the process was far from complete. The interactions between extractant on the one hand and solutes and matrix on the other were clearly not well understood. No good guidelines were available for method development in SFE. Much more knowledge is now available, but a lot of work still has to be carried out to make this knowledge available to the analysts in the laboratory. Methodologies, rules and software packages will have to be designed that help the analysts in developing and translating methods for and in SFE. Without this, SFE will continue to suffer from its 'black magic' tag.

Recently a number of new sample preparation techniques have been introduced. Accelerated Solvent Extraction (ASE) is a technique that was developed to overcome the most important disadvantages of SFE, such as matrix-dependency of extraction yields and restrictor-plugging. In part, ASE is a competitor to SFE, in part the techniques are complementary. Extractions using superheated and supercritical water have been shown to yield pleasing results, but many instrumental problems will have to be overcome before this technique is ready to leave the (academic) research laboratories and enter the 'real world'. One thing that is clear is that SFE inspired the research in these new and interesting areas.

Many of the attractive advantages of SFE are basically advantages of the fact that CO_2 is used as the extraction fluid. The range of application of CO_2 is, unfortunately, limited owing to the nonpolar nature of this solvent. SFE is clearly not a technique that can solve all extraction problems, neither is ASE or any other single extraction technique. Both methods are merely tools in the toolkit of the chromatographer. It is one of the difficult tasks of the analytical chemist to select the sample preparation technique best-suited for the problem at hand. The more tools there are in the toolkit, the larger the chances of finding a sample preparation technique that offers the desired characteristics.

A serious source of concern for the future of SFE is the limited availability of commercial instrumentation. Driven by a strong customer demand and commercial interests, instrument manufacturers released SFE instruments that were, in our view, not ready for the market. Before any technique can be fully accepted, it should be capable of generating reproducible results. This is clearly not the case in SFE. Flow rate control and extract collection will have to be more reliable and reproducible for SFE to become more widely applicable. With a number of instrument manufacturers already withdrawing from the market, this should leave a niche market for a limited number of suppliers providing instruments that no longer suffer from these problems. With this instrumentation, and armed with the significantly improved knowledge of the basic principles and the possibilities and limitations of SFE, the experienced analytical chemist will learn to appreciate SFE as an interesting method for sample preparation. A method with its own unique strengths and weaknesses.

References

Adasoglu, N., Dicer, S. and Bolat, E. (1994) Supercritical fluid extraction of essential oil from Turkish lavender flowers. *J. Supercrit. Fluids*, **7** 93-99.

Andersen, M.R., Swanson, J.T., Porter, N.L. and Richter, B.E. (1989) Supercritical fluid extraction as a sample introduction method for chromatography. *J. Chromatogr. Sci.*, **27** 371-77.

Ashraf-Khorassani, M. and Levy, J.M. (1990) Quantitative analysis of polymer additives in low density polyethylene using supercritical fluid extraction/supercritical fluid chromatography. *J. High Resolut. Chromatogr.*, **13** 742-47.

Ashraf-Khorassani, M., Combs, M.T. and Taylor, L.T. (1995) Effect of moisture on supercritical fluid extraction of polynuclear aromatic hydrocarbons and phenols from soil using automated extraction. *J. High Resolut. Chromatogr.*, **18** 709-12.

Barnabas, I.J., Dean, J.R., Tomlinson, W.R. and Owen, S.P. (1995) Experimental design approach for the extraction of polycyclic aromatic hydrocarbons from soil using supercritical carbon dioxide. *Anal. Chem.*, **67** 2064-69.

Bartle, K.D., Clifford, A.A. and Shilstone, G.F. (1989) Prediction of solubility for tar extraction by supercritical carbon dioxide. *J. Supercrit. Fluids*, **1** 30-34.

Bartle, K.D., Clifford, A.A., Hawthorne, S.B., Langenfeld, J.J., Miller, D.J. and Robinson, R.J. (1990) A model for dynamic extraction using a supercritical fluid. *J. Supercrit. Fluids*, **3** 143-49.

Berg, B.E., Hansen, E.M., Gjorven, S. and Greibrokk, T. (1993) On-line enzymatic reaction extraction, and chromatography of fatty acids and triglycerides with supercritical carbon dioxide. *J. High Resolut. Chromatogr.*, **16** 358-63.

Bowardt, S. and Johansson, B. (1994) Analysis of PCBs in sulfur-containing sediments by off-line supercritical fluid extraction and HRGC-ECD. *Anal. Chem.*, **66** 667-73.

Braybrook, J.H. and Mackay, G.A. (1992) Supercritical fluid extraction of polymer additives for use in biocompatibility testing. *Polymer Int.*, **27** 157-64.

Burford, M.D., Hawthorne, S.B., Miller, D.J. and Braggins, T. (1992) Comparison of methods to prevent restrictor plugging during off-line supercritical fluid extraction. *J. Chromatogr.*, **609** 321-32.

Burford, M.D., Hawthorne, S.B. and Miller, D.J. (1993) Extraction rates of spiked versus native PAHs from heterogeneous environmental samples using supercritical fluid extraction and sonication in methylene blue. *Anal. Chem.*, **35** 1497-505.

Burgess, A.N. and Jackson, K. (1992) The removal of carbon tetrachloride from chlorinated polyisoprene using carbon dioxide. *J. Appl. Polym. Sci.*, **46** 1395-99.

Chambers, R.D. and James, S.R. (1979) *Comprehensive Organic Chemistry*, (ed. J.F. Stoddard), Wheaton & Co. Ltd, Exeter, UK, Vol. 1, p. 525.

Chester, T.L., Pinkston, J.D. and Raynie, D.E. (1992) Supercritical fluid chromatography and extraction. *Anal. Chem.*, **64** 153R-170R.

Cotton, N.J., Bartle, K.D., Clifford, A.A. and Dowle, C.J. (1993a) Rate and extent of supercritical fluid extraction of cyclic trimer from poly(ethylene terephthalate) at elevated temperatures. *J. Chromatogr. Sci.*, **31** 157-61.

Cotton, N.J., Bartle, K.D., Clifford, A.A. and Dowle, C.J. (1993b) Rate and extent of supercritical fluid extraction of additives from polypropylene: diffusion, solubility and matrix effects. *J. Applied Polym. Sci.*, **48** 1607-19.

Crompton, T.R. (1984) *The Analysis of Plastics*, Pergamon Press, Oxford, UK.

Dean, J.R. (ed.) (1993) *Application of Supercritical Fluids in Industrial Analysis*, CRC Press, Inc., Boca Raton, Florida, USA, p. 52.

Dobbs, J.M., Wong, J.M., Lahiere, R.J. and Johnston, K.P. (1987) Modification of supercritical fluid phase behavior using polar cosolvents. *Ind. Eng. Chem. Res.*, **26** 56-65.

Eller, F.J. and King, J.W. (1996) Determination of fat content in foods by analytical SFE. *Semin. Food Anal.*, **1** 145-62.

Engelhardt, H., Zapp, J. and Kolla, P. (1991) Sample preparation by supercritical fluid extraction in environmental food and polymer analysis. *Chromatographia*, **32** 527-37.

Fields, J.A. (1997) Coupling chemical derivatization reactions with supercritical fluid extraction. *J. Chromatogr.*, **785** 239-49.

Francis, A.W. (1954) Ternary systems of liquid carbon dioxide. *J. Phys. Chem.*, **58** 1099-114.

Furton, K.J. and Lin, Q. (1992) The dependence of sorbent/analyte type on observed differences in supercritical fluid extraction efficiencies employing extraction vessels of different dimensions. *Chromatographia*, **34** 185-87.

Furton, K.J. and Lin, Q. (1993) Variation in the supercritical fluid extraction of polychlorinated biphenyls as a function of sorbent type, extraction cell dimension and fluid flow rate. *J. Chromatogr. Sci.*, **31** 201-206.

Ganzler, K., Bati, J. and Valko, K. (1986) Microwave extraction: a novel sample preparation method for chromatography. *J. Chromatogr.*, **371** 299-306.

Gere, D.R., Knipe, C.R., Castelli, P., Hedrick, J., Randall, L.G., Schulenberg-Schnell, H., Schuster, R., Doherty, L., Orolin, J. and Lee, H.B. (1993) Bridging the automation gap between sample preparation and analysis: an overiew of SFE, GC, GC-MS and HPLC applied to environmental samples. *J. Chromatogr. Sci.*, **31** 246-58.

Giddings, J.C., Myers, M.Y., McLaren, L. and Keller, R.A. (1968) High-pressure gas chromatography of nonvolatile species. *Science*, **162** 67-73.

Hale, R.C. and Gaylor, M.O. (1995) Determination of PCBs in fish tissues using supercritical fluid extraction. *Environ. Sci. Technol.*, **29** 1043-47.

Hawthorne, S.B. and Miller, D.J. (1986) Extraction and recovery of organic pollutants from environmental solids and Tenax-GC using supercritical carbon dioxide. *J. Chromatogr. Sci.*, **24** 258-64.

Hawthorne, S.B., Miller, D.J. and Krieger, M.S. (1989a) Coupled SFE-GC: a rapid and simple technique for extracting, identifying and quantifying organic analytes from solids and sorbent resins. *J. Chromatogr. Sci.*, **27** 347-54.

Hawthorne, S.B., Krieger, M.S., Miller, D.J. and Mathiason, M.B. (1989b) Collection and quantitation of methoxylated phenol tracers for atmospheric pollution from residential wood stoves. *Environ. Sci. Technol.*, **23** 470-75.

Hawthorne, S.B., Miller, D.J. and Langenfeld, J.J. (1990) Quantitative analysis using directly coupled supercritical fluid extraction-gas chromatography (SFE-GC) with a conventional split/splitless injection port. *J. Chromatogr. Sci.*, **28** 2-8.

Hawthorne, S.B. (1990) Analytical scale supercritical fluid extraction. *Anal. Chem.*, **62** 633A-642A.

Hawthorne, S.B., Langenfeld, J.J., Miller, D.J. and Burford, M.D. (1992) Comparison of supercritical $CHClF_2$, N_2O and CO_2 for the extraction of polychlorinated biphenyls and polycyclic aromatic hydrocarbons. *Anal. Chem.*, **64** 1614-22.

Hawthorne, S.B., Miller, D.J., Burford, M.D., Langenfeld, J.J., Eckert-Tilotta, S. and Louie, P.K. (1993) Factors controlling quantitative supercritical fluid extraction of environmental samples. *J. Chromatogr.*, **642** 301-17.

Hedrick, J.L. and Taylor, L.T. (1989) Quantitative supercritical fluid extraction/chromatography of a phosphate from aqueous media. *Anal. Chem.*, **61** 1986-88.
Hedrick, J.L. (1992) Pittsburgh Conference 1992, Abstract number 632.
Hierro, M.T.G. and Santa-Maria, G. (1992) Supercritical fluid extraction of vegetable and animal fats with carbon dioxide: a mini review. *Food Chem.*, **45** 189-92.
Howard, A.L., Yoo, W.J., Taylor, L.T., Schweighardt, F.K., Emery, A.P., Chesler, S.N. and MacChrehan, W.A. (1993) Supercritical fluid extraction of environmental analytes using trifluoromethane. *J. Chromatogr. Sci.*, **31** 401-408.
Hunt, T.P., Dowle, C.J. and Greenway, G. (1991) Analysis of poly(vinyl chloride) additives by supercritical fluid extraction and supercritical fluid chromatography. *Analyst*, **116** 1299-304.
Janda, V., Mikesova, M. and Vejrosta, J. (1997) Direct supercritical fluid extraction of water-based matrices. *J. Chromatogr. A*, **733** 35-40.
Janssen, H.-G., Schoenmakers, P.J. and Cramers, C.A. (1989) A fundamental study of the effects of modifiers in supercritical fluid chromatography. *J. High Resolut. Chromatogr.*, **123** 645-55.
Janssen, H.-G., Schoenmakers, P.J. and Cramers, C.A. (1991) Mobile and stationary phases for SFC: effects of using modifiers. *Microchim. Acta.*, **11** 337-46.
Jinno, K. (ed.) (1992) *Hyphenated Systems in Supercritical Fluid Chromatography and Extraction*, Elsevier, Amsterdam, The Netherlands.
Jordan, L., Taylor, L.T., Seemuth, P.D. and Miller, R.J. (1997) Analysis of additives and monomers in nylon and polystyrene. *Textile Chemist and Colorist*, **29** 25-32.
King, J.W., Johnson, J.H. and Friedrich, J.P. (1989) Extraction of fat tissue from meat products with supercritical carbon dioxide. *J. Agric. Food Chem.*, **37** 951-54.
King, J.W. (1989) Fundamentals and applications of supercritical fluid extraction in chromatographic science. *J. Chromatogr. Sci.*, **27** 355-64.
King, J.W. (1990) Applications of capillary supercritical fluid chromatography-supercritical fluid extraction to natural products. *J. Chromatogr. Sci.*, **28** 9-14.
King, J.W. and Friedrich, J.P. (1990) Quantitative correlations between solute molecular structure and solubility in supercritical fluids. *J. Chromatogr.*, **517** 449-58.
Krunic, R.T., Holla, S.J. and Reid, C.R. (1981) Solubility of solids in supercritical carbon dioxide and ethylene. *J. Chem. Eng. Data*, **26** 47-51.
Kueppers, S. (1992) The use of temperature variation in supercritical fluid extraction of polymers for the selective extraction of low molecular weight components from poly(ethylene terephthalate). *Chromatographia*, **33** 434-40.
Langenfeld, J.J., Burford, M.D., Hawthorne, S.B. and Miller, D.J. (1992) Effects of collection solvent parameters and extraction cell geometry on supercritical fluid extraction efficiencies. *J. Chromatogr.*, **594** 297-307.
Langenfeld, J.J., Hawthorne, S.B., Miller, D.J. and Pawliszyn, J. (1993) Effects of temperature and pressure on supercritical fluid extraction efficiencies of polycyclic aromatic hydrocarbons and polychlorinated biphenyls. *Anal. Chem.*, **65** 338-44.
Lee, M.L. and Markides, K.E. (eds.) (1990) *Analytical Supercritical Fluid Chromatography and Extraction*, Chromatography Conference, Provo, UT, USA, 1990.
Lehotay, S.J. (1997) Supercritical fluid extraction of pesticides in food. *J. Chromatogr. A*, **785** 289-312.
Leichter, E., Strode III, J.T.B., Taylor, L.T. and Schweighardt, F.K. (1996) Effect of helium headspace carbon dioxide cylinders on packed-column supercritical fluid chromatography. *Anal. Chem.*, **68** 894-98.
Lembke, P. and Engelhardt, H. (1993) Development of a new SFE method for rapid determination of total fat content of food. *Chromatographia*, **35** 509-16.
Liu, H. and Wehmeyer, K.R. (1992) Solid phase extraction with supercritical fluid elution as a sample preparation technique for the ultratrace analysis of flavone in blood plasma. *J. Chromatogr.*, **577** 61-67.

Lopez-Avilla, V., Young, R., Kim, R. and Beckert, W.F. (1995) Accelerated extraction of organic pollutants using microwave energy. *J. Chromatogr. Sci.*, **33** 481-84.

Lou, X., Janssen, H.-G. and Cramers, C.A. (1995a) Correlation of supercritical fluid extraction recoveries with supercritical fluid chromatographic retention data: a fundamental study. *J. High Resolut. Chromatogr.*, **18** 483-89.

Lou, X., Janssen, H.-G. and Cramers, C.A. (1995b) Investigation of parameters affecting the supercritical fluid extraction of polymer additives from polyethylene. *J. Microcol. Sep.*, **7** 303-17.

Lou, X., Janssen, H.-G. and Cramers, C.A. (1996) Effects of modifier addition and temperature variation in SFE of polymeric materials. *J. Chromatogr. Sci.*, **34** 282-90.

Lou, X. (1997) Thesis, Eindhoven University of Technology.

Lou, X., Janssen, H.-G. and Cramers, C.A. (1997) Temperature and pressure effects on solubility in supercritical carbon dioxide and retention in supercritical fluid chromatography. *J. Chromatogr. A.*, **758** 57-65.

Maguire, K.L. and Denyszyn, R.B. (1988) *Modern Supercritical Fluid Chromatography*, (ed. C.M. White) Huethig, Heidelberg, Germany, pp. 45-73.

Marin, M.L., Jiminez, A. and Vilaplana, J. (1996) Analysis of poly(vinyl chloride) additives by supercritical fluid extraction and gas chromatography. *J. Chromatogr.*, **750** 183-90.

McNally, M.E.P. and Wheeler, J.R. (1988) Increasing extraction efficiency in supercritical fluid extraction from complex samples: predicting extraction efficiency of diuron and linuron in supercritical fluid extraction using supercritical fluid chromatographic retention. *J. Chromatogr.*, **447** 53-63.

McNally, M.E.P. (1995) Advances in environmental SFE. *Anal. Chem.*, **67** 308A.

Meyer, A., Kleibohmer, W. and Camman, K. (1993) SFE of PAHs from soils with a high carbon content and analyte collection via combined liquid/solid trapping. *J. High Resolut. Chromatogr.*, **16** 491-94.

Michaels, A.S. and Bixler, H.J. (1961) Solubility of gases in polyethylene. *J. Polymer Sci.*, **1** 393-412.

Miller, D.J. and Hawthorne, S.B. (1995) Determination of solubilities of organic solutes in supercritical CO_2 by on-line flame ionization detection. *Anal. Chem.*, **67** 273-79.

Moore, W.N. and Taylor, L.T. (1995) Solid phase trapping mechanisms involved in the supercritical fluid extraction of digitalis glycosides with modified carbon dioxide. *Anal. Chem.*, **67** 2030-36.

Mulcahey, L.J. and Taylor, L.T. (1992) Collection efficiency of solid surface and sorbent traps in supercritical fluid extraction with modified carbon dioxide. *Anal. Chem.*, **64** 2352-58.

Ong, C.P., Ong, H.M., Li, S.F.Y. and Lee, H.K. (1990) The extraction of cholesterol from solid and liquid matrices using supercritical CO_2. *J. Microcol. Sep.*, **2** 69-73.

Pariente, G.L., Pentoney, S.L., Griffiths, P.R. and Shafer, K.H. (1987) Computer-controlled pneumatic amplifier pump for supercritical fluid chromatography and extractions. *Anal. Chem.*, **59** 803-13.

Pawliszyn, J. (1993) Kinetic model of supercritical fluid extraction. *J. Chromatogr. Sci.*, **31** 31-37.

Porter, N.L., Rynaski, A.F., Campbell, E.R., Saunders, M., Richter, B.E., Swanson, J.T., Nielsen, R.B. and Murphy, B.J. (1992) Studies of linear restrictors and analyte collection via solvent trapping after supercritical fluid extraction. *J. Chromatogr. Sci.*, **30** 367-73.

Pratte, T.S., Guigard, S., Zytner, R.G. and Silver, W.H. (1996) Influence of water on supercritical fluid extraction. *Proc. Ind. Waste Conf. 1996*, Ann Arbor Press, MI, USA, 51, pp. 95-103.

Randall, L.G. (1982) The present status of dense (supercritical) gas extraction and dense gas chromatography: impetus for DGC/MS development. *Sep. Sci. Technol.*, **17** 1-118.

Raymer, J.H. and Pellizzari, E.D. (1987) Toxic organic compound recoveries from 2,6-diphenyl-*p*-phenylene oxide porous polymers using supercritical carbon dioxide and thermal desorption methods. *Anal. Chem.*, **59** 1043-48.

Rein, J., Cork, C.M. and Furton, K.J. (1991) Factors governing the analytical supercritical fluid extraction and supercritical fluid chromatographic retention of polycyclic aromatic hydrocarbons. *J. Chromatogr.*, **545** 149-60.

Richard, M. and Campbell, R.M. (1991) Comparison of supercritical fluid extraction, Soxhlet and sonication methods for the determination of priority pollutants in soil. *LC-GC*, **4** 358-64.

Richter, B.E., Jones, B.A., Ezzell, J.L., Porter, N.L., Avdalovic, N. and Pohl, C. (1996) Accelerated solvent extraction: a technique for sample preparation. *Anal. Chem.*, **68** 1033-39.

Sandra, P., Kot, A. and David, F. (1994) Supercritical fluids in separation science: trends and concerns. In *Proceedings of Sixteenth International Symposium on Capillary Chromatogr.*, (ed. P. Sandra), Riva del Garda, Italy, 1994, pp. 1515-19.

Schmidt, S., Blomberg, L. and Wannmann, T. (1989) Analysis of volatiles in polymers. Part II. Supercritical fluid extraction/open tubular GC-MS. *Chromatographia*, **28** 400-11.

Shilstone, G.F., Raynor, M.W., Bartle, K.D., Clifford, A.A., Davids, I.L. and Jafar, S.A. (1990) Coupled supercritical fluid extraction/chromatography of polycyclic aromatic hydrocarbons: correlation with solubility calculations. *Polycyc. Aromat. Comp.*, **1** 99-108.

Smith, R.D., Udseth, H.R., Wright, B.W. and Yonker, C.R. (1987) Solubilities in supercritical fluids: the application of chromatographic measurement methods. *Sep. Sci. Technol.*, **22** 1065-86.

Spell, H.L. and Eddy, R.D. (1960) Determination of additives in polyethylene by adsorption spectroscopy. *Anal. Chem.*, **32** 1811-21.

Stahl, E., Schilz, S., Schutz, E. and Willing, E. (1978) A quick method for the microanalytical evaluation of the dissolving power of supercritical gases. *Angew. Chem. Int. Ed. Engl.*, **17** 731-38.

Suto, K., Kakinuma, S., Ito, Y., Sagara, K., Iwasaki, H. and Itokawa, H. (1997) Determination of berberine and palmatine in *Phellodendri cortex* using ion-pair supercritical fluid chromatography on-line coupled with ion-pair supercritical fluid extraction by on-column trapping. *J. Chromatogr.*, **786** 371-76.

Swami, K., Narang, A.S. and Raijnders, S. (1997) Determination of chlordane and chlorpyrifos in ambient air at low nanogram per cubic meter levels by supercritical fluid extraction. *J. AOAC Int.*, **80** (1) 74-78.

Temelli, F. (1992) Extraction of triglycerides and phospholipids from Canola with supercritical carbon dioxide and ethanol. *J. Food Sci.*, **57** 440-57.

Thiebaut, D., Chervet, J.P., Vannoort, R.W., De Jong, G.J., Brinkman, U.A.Th. and Frei, R.W. (1989) Supercritical fluid extraction of aqueous samples and on-line coupling to supercritical fluid chromatography. *J. Chromatogr.*, **477** 151-59.

Valcarcel, M. and Tena, M.T. (1997) Applications of supercritical fluid extraction in food analysis. *Fres. J. Anal. Chem.*, **358** 561.

Vejrosta, J., Ansorgova, A., Mikesova, M. and Bartle, K.D. (1994) Sensitivity enhancement in dynamic 'off-line' supercritical fluid extraction. *J. Chromatogr.*, **659** 209-12.

Venema, A., van de Ven, H.J.F.M., David, F. and Sandra, P. (1993) Supercritical fluid extraction of Nylon-6: an investigation into the factors affecting the efficiency of extraction of caprolactam and its oligomers. *J. High Resolut. Chromatogr.*, **16** 522-24.

Weast, R.C. (1984) (ed), *CRC Handbook of Chemistry and Physics*, 62nd ed, Chemical Rubber Company, Boca Raton, Florida, USA.

Wheeler, J.R. and McNally, M.E. (1989) Supercritical fluid extraction and chromatography of representative agricultural products with capillary and microbore columns. *J. Chromatogr. Sci.*, **27** 534-39.

Wright, B.W., Frye, S.R., McMinn, D.G. and Smith, R.D. (1987a) On-line supercritical fluid extraction-capillary gas chromatography. *Anal. Chem.*, **59** 640-44.

Wright, B.W., Wright, C.W., Gale, R.W. and Smith, R.D. (1987b) Analytical supercritical fluid extraction of adsorbent materials. *Anal. Chem.*, **59** 38-44.

Zhao, S., Wang, R. and Yang, G. (1995) A method for measurement of solid solubility in supercritical carbon dioxide. *J. Supercrit. Fluids*, **8** 15-19.

6 Pressurized fluid extraction (PFE) in organic analysis

John L. Ezzell

6.1 Introduction

In order to assess the organic analyte content of a solid sample, the compounds of interest must be extracted into the liquid state for introduction into quantifying instrumentation. In recent years, there has been a trend toward a reduction in time and in the volume of hazardous solvents used in these sample extraction techniques. The volume of organic solvents required to extract these sample types can compose the largest single source of waste for the analytical laboratory. Alternatives to traditional extraction techniques, such as Soxhlet extraction, have been introduced and have met with varying degrees of acceptance in the sample preparation laboratory. Automated Soxhlet, supercritical fluid and microwave-assisted extraction techniques have all been developed in an attempt to solve the challenges of sample preparation. In each of these approaches, temperature is used as a fundamental parameter.

Pressurized fluid extraction (PFE) is a technique developed for the extraction of solid and semi-solid samples, which uses liquid solvents at temperatures above their atmospheric boiling points. By taking advantage of the increase in analyte solubility and solvent strength which accompany an increase in temperature, PFE is able to increase the efficiency of the extraction process. This results in extractions performed in less time and using less solvent than traditional methods, such as Soxhlet and sonication extraction. With PFE, most extractions are performed in the 75–150°C range. In order to use common organic solvents in this temperature range, pressure is applied to maintain the solvents in their liquid state. A typical extraction is performed in 15 min, using 15–45 ml of solvent for sample sizes ranging 1–30 g. Since its development and commercialization as Accelerated Solvent Extraction (ASE®; Dionex Corp.) (Richter et al., 1995), PFE has gained rapid acceptance in the environmental, polymer, food and pharmaceutical industries.

6.2 Basic principles

In order to extract an organic analyte from a solid matrix into a liquid solvent, there are interactions which must be overcome and interactions which must be formed. Firstly, the interaction between the solute molecule (target analyte) and its neighbouring solute and/or matrix molecules must be broken. This force can be attributed to hydrogen bonding, Van derWaals interactions and dipolar interactions among the molecules. The energy required to break these interactions can be called the lattice energy, the heat of sublimation or the heat of vaporization. This value will generally increase with the polarity of the solute molecule. Secondly, once the solute is free from the matrix, it must be solubilized into the liquid solvent. Initially, this requires the breaking of interactions between solvent molecules. This can occur in two ways: (i) The interactions of the solvent must be weakened to the point that it can 'wet' the solid surface of the matrix; this allows the solvent to exert a force in disrupting the solute-matrix interaction described above. (ii) The solvent network is weakened to the point that the solute can be accommodated into the solvent environment; the energy required for this process is largest in polar solvents, and can be estimated by reference to such parameters as surface tension and boiling point.

Once solubilized into the solvent, the solute now forms a new interaction with the solvent molecules as a stable solvent cage is formed around it. The stability of this interaction results in a release of energy, which is largest when the interaction of the solute and solvent is more polar. In order to achieve maximum solubility in the shortest period of time, the energies described above, which act as barriers to solvation, must be overcome. Increasing the temperature of the process should act to increase the rate of the steps required for extraction to occur. An increase in temperature will act to decrease the magnitude of hydrogen bonding and dipolar interactions occurring on the high energy surfaces of the matrix. Thermal energy will also assist in disrupting stable lattice structures and impart a degree of kinetic energy to the solutes. Furthermore, increased temperature will decrease the surface tension of the liquid solvent (as well as its viscosity), and enhance diffusion. These factors will, in turn, enhance 'wettability', or the ability of a solvent molecule to free itself from its solvent environment and associate with the solid surface of the matrix. In addition, a general decrease in solvent-solvent interactions will more easily accommodate solute molecules into the solvent environment.

Through the application of increased temperature, PFE is able to maximize solubilization by overcoming the energy barriers associated with extraction, and thereby increase the efficiency of the process.

Pressure is a required component of PFE, since most of the organic extraction solvents used have atmospheric boiling points lower than the temperatures at which they are normally used in this technique. Increased pressure can also have a positive role in the extraction process by forcing the solvent into matrix pores and increasing the collision probability of solute and solvent molecules.

6.3 Instrumentation

A schematic diagram of a PFE system is presented in Figure 6.1. The PFE extraction procedure consists of a combination of dynamic and static flow of the solvent through a heated extraction cell containing the sample. These cells must be capable of safely withstanding the pressure requirements of the system, and are normally constructed of stainless steel, with frits in the end caps to allow the passage of solvent whilst retaining the solid sample within. The pore size of the frit should not allow passage of the matrix particles (5–10 µm is typical). Disposable filters may be used in the cell outlets to avoid compaction of fine particles on the frit surface, which may impede the solvent flow. The sample cell is interfaced to the solvent flow path, where it is filled with solvent. It is important to ensure that all of the void volume has been filled with solvent in order to ensure good contact between the sample matrix and the solvent, and to avoid possible analyte oxidation, which may occur in the presence of oxygen at elevated temperatures.

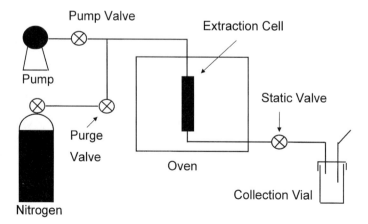

Figure 6.1 Schematic representation of a pressurized fluid extraction (PFE) system.

The sample cell is then heated by contact with a stable heat source. Heating the cell prior to introduction of the solvent can result in the loss of volatile compounds (Richter *et al.*, 1996). In order to accurately assess the effect of the extraction temperature, the inside of the cell must be allowed to reach thermal equilibrium with the heating source. The time required for heating will vary depending on the degree of contact of the cell with the heat source, the heat transfer characteristics of the cell walls, and the temperature set. For a 100°C extraction, 5 min allows sufficient time to heat a sample in a stainless steel cell with a 2 mm wall thickness. In order to maintain the extraction solvents in their liquid state, a pressure source must be applied. The pressure of the system must be above the threshold required to maintain the liquid state of the solvent at the set temperature and must be able to move the solvent through the sample cell in a reasonable time period. This is normally accomplished with a high-performance liquid chromatography (HPLC)-type pump, which can maintain a constant fluid pressure of 1000–3000 psi (6.9–20.7 MPa). With the sample cell interfaced to the system flow, the basic configuration resembles a HPLC system, using a pump to force the liquid solvent through a packed bed (granular sample).

When heating a liquid solvent in a closed cell, the solvent will expand and result in an increase in pressure. This increase must be regulated without the loss of solvent from the cell (which contains target analytes). In order to accomplish this, a valve is positioned between the extraction cell and the collection vial. This valve is pulsed at intervals during the heating process, allowing small volumes of fluid to escape from the cell into the vial. Due to the constant pressure applied by the pump, an additional amount of fresh solvent is introduced to the head of the cell as the valve pulses, maintaining the cell volume. Once thermal equilibrium has been reached, the sample cell is maintained at temperature for an additional time period, which is usually equal to the heating time or approximately 5 min. During this static phase, analyte diffusion from the matrix into the solvent occurs.

Following this static hold step, the outlet valve is opened and a measured volume of fresh solvent, usually 50–70% of the cell volume, is allowed to flush over the sample, expelling the previous volume into the collection vial. Finally, compressed nitrogen is used to force all of the solvent from the lines and cell, into the vial. It is important that all of the solvent used in the extraction be collected for analysis. Since the final stage of the process involves blowing nitrogen over the extract, some volatilization of solvent may occur, and the system can be vented to a fume hood or other device if required. The collection vials normally used are standard 40 or 60 ml vials suitable for volatile organics analysis, sealed with a Teflon®-coated septa (Teflon is a registered trademark of

Dupont). This allows the extracts to be collected and maintained in a sealed, inert environment (under a nitrogen blanket) in order to prevent sample loss while waiting for quantification. When working with potentially flammable solvents near a heated source, care must be taken to avoid solvent spills, and pressurized fittings should be checked regularly for integrity. PFE is normally accomplished with a fluid flow from the top to the bottom of the cell, as this facilitates purging of the liquid with the compressed gas source, however, flow direction is not critical. Due to the high operating temperatures of the system, solvents with unusually low autoignition points, such as carbon disulphide (CS_2), should be avoided.

6.4 PFE methods development

6.4.1 Sample preparation

Sample preparation is an essential part of every solvent-based extraction procedure. While many sample types can be efficiently extracted without any pretreatment, other samples will require some manipulation for an efficient extraction to occur. As with Soxhlet, the ideal sample for extraction is a dry, finely-divided solid. Whatever can be done to make the sample approach this definition will have a positive impact on the results of the extraction. In general, the same sample preparation that is carried out prior to Soxhlet or sonication extraction should be performed prior to extraction by PFE.

6.4.2 Grinding

For an efficient extraction to occur, the solvent must be able to make contact with the target analytes. The greater the surface area that can be exposed in a sample, the faster this will occur. Samples with large particle sizes should be ground prior to extraction. The minimum particle size required may differ based on the sample type, but should generally be smaller than 1.0 mm. Grinding can be accomplished with a conventional mortar and pestle, or with electric grinders and mills. Since quantitative transfer of ground material can be difficult, it is recommended that a large, representative sample be ground, and weighed portions of the ground sample be used for extraction. Polymer samples should be in a ground state for an efficient extraction of additive compounds. This is best accomplished by using a liquid nitrogen cryogrinder.

6.4.3 Dispersing

It is possible that the interaction of sample particles may prevent efficient extraction. In such cases, dispersing the sample with an inert material, such as sand or diatomaceous earth (DE), will assist in the extraction process. Dispersion will also assist the extraction process when dealing with samples which tend to compact in the extraction cell outlet. These samples normally contain very small particles ('fines'), which can adhere tightly to each other under pressure. The use of sea sand is not recommended as a dispersing agent in PFE. This material can contain very small particles which can block system tubing. When using a dispersing agent, it is a good idea to periodically run blank extractions of the material to verify its cleanliness.

6.4.4 Drying

Many environmental samples contain water which can act as a shield, preventing nonpolar organic solvents from reaching the target analytes. While the use of more polar solvents (acetone, methanol) or solvent mixtures (hexane/acetone, methylene chloride/acetone) can be used to assist in the extraction of wet samples, drying of the sample prior to extraction is the most efficient way to handle these sample types. Drying is normally accomplished by direct addition of a material such as sodium sulphate, DE or cellulose. The choice of drying agent depends on the preference of the user. While sodium sulphate works well for soil and sediment samples, pelleted DE is a good choice for wet tissue samples. Cellulose may be used for very wet, soft matrices, such as fruits and vegetables. The use of magnesium sulphate is not recommended in PFE, due to its potential for melting at higher temperatures. Sodium sulphate should not be used with methanol or other polar solvents, as it will be solubilized at elevated temperatures. Oven-drying and freeze-drying are other viable alternatives for drying samples prior to extraction, however, the recovery of volatile compounds may be compromised by these procedures.

The following ratios have been applied to various sample types for PFE extraction: samples which appear dry, 4 g sample to 1 g DE or 4 g Na_2SO_4; and samples which appear wet, 4 g sample to 2 g DE or 8 g Na_2SO_4.

The sample and the drying or dispersing agent should be mixed thoroughly in a small vial, beaker or mortar, and then added to the extraction cell. For quantitative transfer, the mixing vessel can be rinsed with 1–2 ml of the extraction solvent using a Pasteur pipette, and this volume added directly to the extraction cell. Surrogate standards or other internal standard mixtures can be added directly to the extraction cell or collection vial.

The following are examples of sample preparation: wet soil/sediment, 10 g sample + 5 g DE or 20 g sodium sulphate; fish tissue (80% moisture), 3 g sample + 2 g DE or 15 g sodium sulphate; fruits and vegetables, 10 g sample + 5 g DE or 2 g cellulose; creams and lotions, 2 g sample + 3 g DE; and ground polymers, 1–3 g sample + 1–3 g sand.

6.4.5 Extraction parameters

i) Solvent

In order for an efficient extraction to occur, the solvent must be able to solubilize the target analytes, while leaving the sample matrix generally intact. The polarity of the extraction solvent should closely match that of the target compounds. Mixing solvents of different polarities can be used to extract broad range compound classes. Generally, if a particular solvent has been shown to work well in a conventional procedure, it will also work well in PFE. Therefore, the analysts experience with a particular sample type or analyte class should be used. Compatibility with the postextraction analytical technique, the need for extract concentration (solvent volatility) and the cost of the solvent should all be considered. While many PFE methods recommend solvents or solvent mixtures for specific classes of analyte, there may be alternatives which are more suitable for the needs of a particular laboratory. Solvents which exhibit marginal results in ambient conditions may perform adequately under PFE conditions. Most liquid solvents, including water and buffered aqueous mixtures, can be used in PFE. Strong acids (HCl, HNO_3, H_2SO_4) are not recommended for use, due to their ability to react with the stainless steel in most systems. When acidic conditions are required, weaker acids, such as acetic or phosphoric acid, should be used, and are normally added to aqueous or polar solvents in the 1–10% (v/v) range. When using the same solvent system as a traditional method, analyte recovery as well as co-extractable content should be very similar. Therefore, if a traditional extraction approach requires an extract clean-up step prior to quantification, the PFE extract will probably require the same manipulation.

ii) Temperature

Temperature is the most important parameter used in PFE extraction. When developing a new method, it is recommended to start at 100°C, or if the target analytes have a known thermal degradation point to start at 20°C below this level. Most PFE applications operate in the 75–150°C range, with 100°C being the standard temperature for all environmental applications except dioxins (150°C). An example of the effect of temperature is shown for the extraction of total petroleum hydrocarbons (TPH) from a standard reference soil (Table 6.1).

Table 6.1 Effect of temperature on recovery and reproducibility of TPH from a standard reference soil (n = 5)

Temperature °C	Recovery %	RSD %
27	81.2	6.0
50	93.2	5.0
75	99.2	2.0
100	102.7	1.0

Abbreviations: TPH, total petroleum hydrocarbons; RSD, relative standard deviation.

Note that, not only does the analyte recovery increase but the reproducibility improves with increased temperature. An increase in temperature will generally have a positive impact on the extraction but only up to the point where analyte or matrix degradation begins to occur. These problems, however, are not generally observed in the normal operating ranges of PFE. For sample matrices which tend to melt at elevated temperatures, cellulose extraction thimbles can be inserted into the extraction cells, in order to facilitate sample loading and removal.

iii) Pressure
The effect of pressure is, firstly, to maintain the solvents as liquids while above their atmospheric boiling points and, secondly, to move the fluids through the system rapidly. The pressures normally used in PFE are well above the thresholds required to maintain the solvents in their liquid states (1000–3000 psi, 6.9–20.7 MPa), so adjustments for changing solvents are not required. Changing the pressure will have very little impact on analyte recovery, and it is not considered a critical experimental parameter. Most PFE extractions are performed between 1000 and 2000 psi (6.9–13.8 MPa), with 1500 psi (10.3 MPa) being the standard value.

iv) Cycles
The use of static cycles was developed in order to introduce fresh solvent during the extraction process, which helps to maintain a favourable extraction equilibrium. This effectively approximates dynamic extraction conditions without the need for troublesome flow restrictors to maintain pressure. When more than one cycle is used in a method, the flush volume is divided by that number. When the first static time is complete, the

divided portion of the flush volume is delivered to the cell, with the 'used' solvent directed to the collection vial. The system then holds the sample and solvent for a second static period. The nitrogen purge step is initiated only after the final static cycle. Since the original flush volume has only been divided, no additional solvent is used for the extraction. Static cycles have proved useful for sample types with a very high concentration of analyte or samples with matrices that are difficult to penetrate. The static time can be adjusted to minimize the total extraction time. For example, three 3 min static cycles can be used in place of one 10 min static step. When low temperature extractions are desired ($<75°C$), multiple static cycles should be used to compensate for the lack of fresh solvent normally introduced during the heat-up step, as the static valve pulses to regulate the pressure.

v) Time

Certain sample matrices can retain analytes bound, encapsulated or otherwise trapped within pores or other structures. Increasing the static time at elevated temperatures can allow these compounds to diffuse into the extraction solvent. The effect of static time should always be explored in conjunction with static cycles, in order to produce a complete extraction in the most efficient way possible.

6.4.6 Method validation

When developing a method for PFE the following approach has proved useful. A representative sample is prepared as outlined above. An extraction cell size which most closely matches the desired sample size is selected. The extraction cell does not need to be filled completely; however, a full cell will use less solvent in the extraction process than a partially filled one. The extraction solvent should be selected using the considerations listed above, although normally the same solvent or solvent mixture used in a traditional liquid extraction method is used. The sample is extracted, starting with the standard PFE conditions as follows: pressure, 1500 psi (10.3 MPa); temperature, 100°C; static time, 5 min; flush volume, 60% of cell volume; purge time, 60 s; and number of cycles, 1.

The same sample is extracted several times in order to assess the efficiency of the method. If there is significant analyte present in the second or third extract, the following parameters should be adjusted (one at a time), and the validation process repeated: (i) increasing the temperature (using 20°C steps); (ii) adding a second or third static cycle; (iii) increasing the static time (using 5 min increments). If these steps do not result in a complete extraction, the sample preparation steps and/or the choice of extraction solvent should be re-examined.

6.4.7 Selectivity

Once an extraction method has been developed and shown to produce adequate recovery of target analytes, the analyst may try to optimize selectivity. Selectivity in extraction can be defined as the ability to extract only the analyte(s) of interest, while leaving other compounds behind. The presence of these co-extractable materials may necessitate a postextraction clean-up step to remove compounds which interfere with chromatographic analysis. Complete selectivity, extraction of solely target compounds, is normally difficult to accomplish due to the closely related solubility characteristics of compounds in a sample matrix. However, by applying knowledge of the sample matrix and its constituents, a degree of selectivity may be achieved. The PFE extraction parameters which most effect selectivity in extraction are choice of solvent and extraction temperature. If, in a given set of conditions, the target compounds are recovered with a high degree of co-extractable materials, a shift in solvent polarity or a reduction in extraction temperature may result in a reduction of the contaminant material. It is also possible to add selective sorbent materials to the extraction cell outlet or the collection vial, to retain unwanted components from the extract flow. This was done using activated alumina to retain unwanted materials from fish tissue extracts (Ezzell *et al.*, 1996b), and by using copper powder to retain sulphur from sediment extracts (Schantz *et al.*, 1997). In these cases, the effect of the sorbent should be tested for retention of target materials, and will be specific for the analyte/co-extractant/solvent system. In some cases, the polarity and solubility of the target compounds may be so closely related to the co-extractables as to make selective isolation at the extraction step improbable.

6.5 Strengths/limitations

Compared to conventional extraction times, ranging 4–48 h in length, PFE extractions are normally performed in 12–20 min per sample. While the decrease in extraction time is favourable for laboratories in general, it can be critical for those industries where data from laboratory analysis are used in feedback control of production cycles and quality control of manufacturing processes. The volume of solvents used in PFE can be 10–20 times less than traditional extraction methods. When factors such as safety and analyst exposure, solvent purchase and disposal costs (including vent hood requirements), are considered, the benefits of PFE can be quite substantial for most laboratories. When compared directly to traditional extraction methods, the recoveries generated by PFE normally equal or slightly exceed the comparative method (Richter *et al.*, 1995; Ezzell *et al.*, 1995; Schantz *et al.*, 1997). PFE has been extensively compared to traditional methods used in the environmental industry, and standard methods are available for most classes of analyte. All of these methods are essentially the same, with the exception of the extraction solvent, which is chosen based on the target analyte class and is generally the same solvent or solvent mixture used in the traditional methods. The ability of PFE to use the same liquid solvents used in traditional extraction methods allows for rapid conversion to this technique, without much effort involved in methods development. Once a PFE method has been developed for a class of analyte, that same method can be successfully applied to a variety of matrix types without adjustment to extraction parameters. This lack of matrix dependency has allowed a very small set of standard methods to be applied to a large range of sample types.

The general approach of PFE is to move a liquid solvent through a sample bed in a heated cell. While this results in an extract collected away from the sample matrix, eliminating the need for postextraction filtering, it also renders the system unsuitable for the extraction of liquid samples. However, PFE can be used to extract solid phase extraction (SPE) or other adsorbent filter materials following extraction of large volume liquid or air samples. Since pressure must be applied and maintained within the cell during the extraction process, sample sizes are limited to that which can fit into the pressure-rated vessels. Normally, the 1–30 g sample range is appropriate for PFE systems.

The increased temperatures used in PFE increase the efficiency of the extraction process. Generally, an increase in extraction temperature will result in an increase in the amount of material extracted from a sample matrix. There is a point, however, at which increased temperature begins to have a negative effect on analyte recovery. This is normally due to thermal degradation or thermally-induced rearrangement of target compounds. Burning or charring of sample matrices can also occur as the temperature is increased. Extensive evaluations and comparisons of PFE have shown that in the normal operating range of 75–150°C, the extraction efficiency is increased without these negative effects. As the temperature is increased towards the 200°C level, however, stability problems with analytes and matrices are more common. It seems unlikely, therefore, that future generation PFE systems will incorporate higher temperature capabilities, and methods development in high temperature extraction (> 200°C) will probably be limited to very specific areas of application. PFE is generally considered an exhaustive extraction technique, and under the appropriate conditions all of the extractable content of a matrix will be solubilized. At the temperatures used in PFE, this will include the moisture content of the sample matrix. If the extraction solvent has a degree of miscibility with water (methanol, acetone, etc.) the analyst may not observe water in the collection vial. However, when nonpolar solvents are used, an aqueous layer may be present. This is normally eliminated from the extract prior to concentration or analysis by addition of sodium sulphate directly to the vial, shaking and decanting.

In contrast to traditional extraction approaches, all of the basic steps of PFE are amenable to automation, freeing the analyst from the labour-intensive nature of most sample preparation protocols. Automated PFE systems can extract up to 24 sample cells, and can have the necessary safety considerations for unattended operation built-in. This level of system operation results, however, in a capital cost higher than traditional glassware-based approaches.

6.6 Areas of application

6.6.1 Environmental

Most of the initial studies and publications involving PFE have related to the environmental area, and PFE has been extensively compared to traditional extraction methods. A two laboratory validation study was conducted in 1994 to compare PFE to Soxhlet and automated Soxhlet extraction of environmental target compounds (Richter et al., 1995; Ezzell et al., 1995). In the first study, soils classified as clay, loam and sand were batch prepared with known amounts of organochlorine pesticides (OCPs) (20 compounds) or semi-volatiles (base neutrals and acids [BNAs]) (56 compounds) at three concentration levels. These soils were extracted by PFE and automated Soxhlet, with seven replicates performed for each soil type and concentration level. The average relative recoveries and reproducibility are summarized in Table 6.2.

Table 6.2 Summary of laboratory validation studies used for the generation of US EPA Method 3545

Compound class	Comparison technique	Relative recovery (%)
Organochlorine pesticides	Automated Soxhlet	97.3
Semi-volatiles (BNAs)	Automated Soxhlet	99.2
Organophosphorus pesticides	Soxhlet	98.6
Chlorinated herbicides	Shake method	113
Polychlorinated biphenyls	Various reference materials	98.2
Polycyclic aromatic hydrocarbons	Various reference materials	105

Abbreviations: EPA, Environmental Protection Agency; BNAs, base neutrals and acids.

As part of this study, the thermally-mediated conversion of DDT to DDE and of endrin to endrin aldehyde was monitored and found to be within acceptable limits. The second study evaluated the extraction efficiency of organophosphorus pesticides (OPPs) (24 compounds), herbicides (Herb) (eight compounds) and polychlorinated biphenyls (PCBs) from the same three soil types at two concentration levels. The comparative methods used for this work were traditional Soxhlet extraction for the OPPs, wrist-shaker extraction for the herbicides, and comparison to standard reference material (SRM) values for the PCBs. Relative recovery data are also presented in Table 6.2. The data which resulted from these studies were submitted to the US Environmental Protection Agency (EPA) and resulted in the generation of US EPA Method 3545, which received final promulgation in June 1997 as part of SW-846 Method Update III (US GPO, 1997).

Additional work has since been performed and PFE technology validated on the same classes of compounds (Hoefler *et al.*, 1995a, b, c; Conte *et al.*, 1997; Fisher *et al.*, 1997; Heemken *et al.*, 1997). A comprehensive evaluation of the extraction of SRMs was performed by the National Institute of Standards and Testing (NIST) in 1997, and concluded that the performance of PFE was equivalent to Soxhlet for the recovery of organics from air particulates, sediment, samples, mussel tissue and fish materials, and that PFE generated better extraction efficiencies for higher molecular weight polycyclic aromatic hydrocarbons (PAHs) in diesel particulate materials (Schantz *et al.*, 1997). In all of these studies, no significant variation from the methods originally published was described. The extraction of PCB congeners from fish tissue, with selective retention of co-extracted interferents was reported in 1996. In this work, the co-extracted materials were trapped on activated alumina and placed in the outlet of the extraction cell, while the extraction efficiency of the PCBs was unaffected (Ezzell *et al.*, 1996b). The use of PFE for the extraction of arochlor 1254 from air filters containing polyurethane foam inserts has also been reported (Ezzell and Richter, 1996a).

TPH and BTEX (benzene, toluene, ethylbenzene, xylene) extraction has been performed, using both IR and gas chromatography (GC) for quantitative analysis. In these cases, the data generated by PFE were shown to be equivalent to the comparative method (Ezzell and Richter, 1996a). In this study, it was shown that due to the closed nature of the PFE system flow path, extraction and retention of volatile compounds is generally quite good. The compound with the lowest boiling point recovered by PFE has been pentane, at the 90% recovery level. The extraction of soils contaminated with organophosphorus hydraulic fluids was reported by David and Seiber (1996). This work concluded that PFE provided extraction efficiencies comparable to SFE and Soxhlet. The Ontario Ministry of Environment and Energy performed a extensive study comparing PFE and Soxhlet for the extraction of dioxins and furans from fly ash, chimney brick, urban dust and sediment samples. Their work concluded that a 19 min PFE using 15–20 ml of toluene at 150°C produced results equivalent to an 18 h Soxhlet using 250 ml of solvent per sample (Richter *et al.*, 1997). Recently, the extraction of explosives from soils and residual pesticides in grain has also been examined (Ezzell, 1998).

A summary of PFE methods used to extract most classes of environmental analytes is presented in Table 6.3. Note that, with the exception of the extraction solvents (chosen on the basis of the target analyte) and the increased temperature used for dioxins, all of the methods are essentially the same, illustrating the lack of matrix dependency attributed to this technique.

6.6.2 *Polymers*

Polymer samples are extracted, not only for analysis of the oligomeric structure but also for the determination of additive compounds included in the formulations to impart specific properties to the materials. Extraction of monomers and oligomers from Nylon 6 and poly-1,4-butylene terephthalate (PBT) was reported by Lou and co-workers (1997). In this study, PFE parameters were optimized and resulted in the use of hexane at 150°C. Results were compared directly to a 30 h Soxhlet extraction and found to be comparable. For efficient extraction of additives from polymers, such as polypropylene, polyethylene, polystyrene or polyvinylchloride, the samples should be in the ground state. Since the goal of PFE is the extraction of the additives and not the polymer matrix itself, more polar solvents are normally used. A 1:1 mixture of acetonitrile and ethyl acetate or a 95:5 mixture of isopropyl alcohol (IPA):cyclohexane has been used to extract butylated hydroxytoluene (BHT), Irganox 1010, Irganox 1076 and Irgafos 168 from ground high density polyethylene (HDPE) at 125°C (Table 6.4) (Carlson *et al.*, 1997). In these extractions, it is possible to go above the melting point of the material; however, this can be problematical since the cooled and hardened material can be difficult to remove from the sample cells. The use of cellulose thimbles inside the extraction cells has proved useful when this occurs. Total extractables from styrene-butadiene rubbers, including naphthanic oils, aromatic oils and organic acids, was performed using IPA at 185°C. Results were comparable to data generated by the American Society of Testing and Materials (ASTM) method D1416,

Table 6.3 Reference table of sample types and general parameters for the pressurized fluid extraction (PFE) method

Analyte class	Matrices	Solvent	Temperature	Pressure (psi)	Static time (min × cycles)	References
Chlorinated pesticides	Soil, sediment, grains, fruits	Hexane/acetone (1/1)	100°C	2000	5 × 1	Richter et al. (1995), Schantz et al. (1997)
Herbicides	Soil	Acetone/MeCl$_2$ (2/1) + 5% H$_3$PO$_4$/water (1/1)	100°C	2000	5 × 1	Ezzell et al. (1995), Ezzell (1997)
Organophosphorus pesticides	Soil	Acetone/MeCl$_2$ (1/1)	100°C	2000	5 × 1	Ezzell et al. (1995)
Semi-volatiles (BNAs)	Soil, sediment, sludge	Acetone/MeCl$_2$ (1/1)	100°C	2000	5 × 1	Richter et al. (1995), Fisher et al. (1997)
PCBs	Soil, sediment, marine tissue, PUFs	Hexane, hexane/acetone (1/1) MeCl$_2$	100°C	1500–2000	5 × 1	Richter et al. (1996), Ezzell et al. (1996b), Schantz et al. (1997), Heempken et al. (1997)
PAHs	Soil, sediment, sludges	Acetone/MeCl$_2$ (1/1)	100°C	1500–200	5 × 1	Richter et al. (1995), Hoefler et al. (1995a,b,c), Schantz et al. (1997), Heempken et al. (1997), Fisher et al. (1997)
TPH	Soils	MeCl$_2$, hexane, PERC, freon	100–200°C	1500–2000	5 × 1	Richter et al. (1996), Ezzell and Richter (1996a)
BTEX	Soils	MeCl$_2$, hexane, PERC, freon	75–100°C	1500–2000	5 × 1	Ezzell and Richter (1996a)
Dioxins/furans	Sediment, dust fly ash, brick	Toluene	150–175°C	1500	5 × 2	Richter et al. (1997)

Analyte	Matrix	Solvent	Temperature	Pressure	Extraction	Reference
Explosives	Soil	MeOH, acetone, acetonitrile	100°C	1500	5 × 1	Ezzell (1998)
Organophosphorus hydraulic fluids	Soil	Methanol	100°C	1500	5 × 1	David and Seiber (1996)
Polymer and polymer additives	Nylon, polypropylene, polyethyelene	Hexane, acetonitrile/ erthyl acetate (1/1)	100–140°C	1500	3 × 3	Lou et al. (1997) Carlson et al. (1997)
Fat	Various foods	Petroleum ether, hexane, hexane/IPA (3/2)	100–125°C	1000–2000	5 × 3	Francis et al. (1997)
Drugs and natural products	Feeds, patches, plants	MeOH, EtOH, acetonitrile	Ambient 100°C	1500	5 × 1	Ezzell et al. (1996c) Peng et al. (1997)

Abbreviations: PCB, polychlorinated biphenyl; PAH, polycyclic aromatic hydrocarbon; TPH, total petroleum hydrocarbons; BNAs, base neutrals and acids; BTEX, benzene, toluene, ethylbenzene, xylene; PUF, polyurethane foam; PERC, perchloroethylene; IPA, isopropyl alcohol.

Table 6.4 Extraction of polymer additives from high density polyethylene

Compound:	BHT	Irganox 1010	Irganox 1076	Irgafos 168
% formulation value recovered	110.9	89.7	87.1	93.3
RSD (%) (n = 5)	15.0	3.1	0.6	2.3

Samples were ground and extracted at 140°C with a 1:1 mixture of acetonitrile and ethyl acetate. Abbreviations: BHT, butylated hydroxytoluene; RSD, relative standard deviation.

which boils the samples in 400 ml ethanol:toluene (70:30) (Carlson et al., 1997).

6.6.3 Foods

The use of PFE in the analysis of food samples has focused on the extraction of residual pesticides, PCBs and herbicides, and the determination of fat content. Residual herbicides have been extracted from grains (Ezzell, 1997) and chlorinated pesticides and PCBs from marine food products (Richter et al., 1996). The extraction of fat from various food matrices by PFE has been compared to traditional techniques, with and without the use of acid hydrolysis. Table 6.5

Table 6.5 Extraction of fat from food matrices by PFE and Soxhlet

Extraction method	Sample	Solvent	Weight % fat
PFE	Sweet cereal	MeOH/CHCl$_3$ (2:1)	11.6
Soxhlet	Sweet cereal	Hexane/IPA (3:2)	10.0–12.0
PFE	Snack cracker	Hexane/IPA (3:2)	1.43
Soxhlet	Snack cracker	Hexane/IPA (3:2)	1.40
PFE	Flour	Petroleum ether	1.37
Soxhlet	Flour	Petroleum ether	1.36

PFE extractions were performed with the solvents indicated at 100–125°C for 20 min. Soxhlet extractions were performed by 4h reflux. Abbreviations: PFE, pressurized fluid extraction; IPA, isopropyl alcohol.

summarizes data obtained from a variety of sample types. The extraction of a NIST standard reference material (SRM 1846) powdered infant formula, concluded that extraction solvent and temperature are important parameters in optimizing extraction conditions. Comparison of gravimetric and fatty acid methyl ester (FAME) chromatographic data from this sample indicated that close correlation to the reported value could be obtained without acid hydrolysis of the sample. Additional comparisons with samples of salad dressing, mayonnaise, cream cheese, processed cheese and peanut butter also showed good correlation with values obtained by traditional techniques with gravimetric analysis (Francis et al., 1997).

6.6.4 Pharmaceuticals and natural products

Extraction and analysis of pharmaceutical preparations is an integral part of the quality assurance system of the pharmaceutical industry. The use of ASE technology in this area offers many advantages over existing methodologies. Many of the methods in use are labour-intensive, involving shaking, decanting and filtering of extracts. Since many of the target analytes are polar in nature and may be temperature sensitive as well, the ability to select from a wide range of extraction solvents and temperatures is essential. Pressurized fluid extraction has been successfully applied to the extraction of drugs from animal feeds using methanol at 100°C, and active components in transdermal patches using ethanol at ambient temperature (Table 6.6A and B) (Ezzell et al., 1996c). In the area of characterization of natural products, the extraction of capsaicinoids and alkaloids from raw materials using methanol at 100°C has been reported, with results comparing favourably to 5 h, reflux-based extraction methods (Peng et al., 1997).

Table 6.6A Recovery of anti-schizophrenic agent from rodent feed

Extraction method	Recovery (g/kg) (% RSD*) 0.2 g/kg feed level	Recovery (g/kg) (% RSD*) 10 g/kg feed level
PFE	0.185 (4.2)	9.68 (4.4)
Wrist-shaker	0.170 (1.3)	9.43 (1.3)

10 g samples extracted by PFE with methanol at 100°C, or wrist-shaker with acetonitrile at ambient temperature (*n = 10). Abbreviations: PFE, pressurized fluid extraction; RSD, relative standard deviation.

Table 6.6B Recovery of nitroglycerin from two sizes of transdermal patch by PFE and sonication

Extraction method	Recovery (mg) (% RSD*) (mg/10 cm² patch)	Recovery (mg) (% RSD*) (mg/20 cm² patch)
PFE	31.4 (1.4)	62.0 (3.9)
Sonication	31.7 (1.5)	64.6 (1.1)

Extractions were performed with ethanol at ambient temperature (*n = 10). Abbreviations: PFE, pressurized fluid extraction; RSD, relative standard deviation.

References

Carlson, R., Clark, J., Ezzell, J. and Joyce, R. (1997) Accelerated solvent extraction of polymer additives. 10th International Symposium on Polymer Analysis and Characterization, Toronto, Ontario, Canada.

Conte, E., Milani, R., Morali, G. and Aballe, F. (1997) Comparison between accelerated solvent extraction and traditional extraction methods for the analysis of the herbicide, difluenican, in soil. *J. Chromatogr. A*, **765** 121-25.

David, M. and Seiber, J. (1996) Comparison of extraction techniques, including supercritical fluid, high-pressure solvent, and Soxhlet, for organophosphorus hydraulic fluids from soil. *Anal. Chem.*, **68** 3038-44.

Ezzell, J., Richter, B., Felix, D., Black, S. and Meikle, J. (1995) A comparison of accelerated solvent extraction with conventional solvent extraction for organophosphorus pesticides and herbicides. *LC/GC*, May, 390-98.

Ezzell, J. and Richter, B. (1996a) Automated sample preparation for environmental laboratories using accelerated solvent extraction. *American Env. Lab.*, February, 16-18.

Ezzell, J., Richter, B. and Francis, E. (1996b) Selective extraction of PCBs from fish tissue using accelerated solvent extraction. *American Env. Lab.*, December, 12-13.

Ezzell, J., Richter, B. and DeCrosta, M. (1996c) Accelerated solvent extraction for preparation of samples of pharmaceutical importance. The Pittsburgh Conference on Analytical Chemistry and Applied Spectroscopy, Chicago, IL, USA, #946.

Ezzell, J. (1997) Accelerated solvent extraction (ASE): the use of SW-846 Method 3545 for automated extraction of environmental samples. *American Env. Lab.*, **10** (2) 24-25.

Fisher, J., Scarlett, M. and Stoit, A. (1997) Accelerated solvent extraction: an evaluation for screening of soils for selected US EPA semivolatile organic priority pollutants. *Environ. Sci. Technol.*, **31** 1120-27.

Francis, E., Knowles, D. and Richter, B. (1997) The determination of fats in food using accelerated solvent extraction (ASE). The Pittsburgh Conference on Analytical Chemistry and Applied Spectroscopy, Atlanta, GA, USA, #29.

Heemken, O., Theobald, N. and Wenclawiak, B. (1997) Comparison of ASE and SFE with Soxhlet, sonication and methanolic saponification for the determination of organic micropollutants in marine particulate matter. *Anal. Chem.*, **69** 2171-80.

Hoefler, F., Ezzell, J. and Richter, B. (1995a) Beschleunigte Losemittel-extraktion (ASE). *LaborPraxis*, March, 64-67.

Hoefler, F., Ezzell, J. and Richter, B. (1995b) Anwendung der ASE in der Umweltanalytik. *LaborPraxis*, April, 58-62.

Hoefler, F., Jensen, D., Ezzell, J. and Richter, B. (1995c) Accelerated solvent extraction of PAH from solid samples with subsequent HPLC analysis. *Chromatographie*, January, 68-71.

Lou, X., Janssen, H. and Cramers, C. (1997) Parameters affecting the accelerated solvent extraction of polymeric samples. *Anal. Chem.*, **69** 1598-1603.

Peng, T., Cao, S., Davis, R. and Johnston, D. (1997) Accelerated solvent extraction of natural products. 111th AOAC International Meeting, September 1997, San Diego, CA, USA, #1102.

Richter, B., Ezzell, J., Felix, D., Roberts, K. and Later, D. (1995) An accelerated solvent extraction system for the rapid preparation of environmental organic compounds in soil. *American Laboratory*, February, 24-28.

Richter, B., Jones, B., Ezzell, J., Porter, N., Avdalovic, N. and Pohl, C. (1996) Accelerated solvent extraction: a technique for sample preparation. *Anal. Chem.*, **68** 1033-39.

Richter, B., Ezzell, J., Knowles, D., Hoefler, F., Mattulat, A., Scheutwinkel, M., Waddell, D., Gobran, T. and Khurana, V. (1997) Extraction of polychlorinated dibenzo-*p*-dioxins and polychlorinated dibenzofurans from environmental samples using accelerated solvent extraction (ASE). *Chemosphere*, **34** 975-87.

Schantz, M., Nichols, J. and Wise, S. (1997) Evaluation of pressurized fluid extraction for the extraction of environmental matrix reference materials. *Anal. Chem.*, **69** 4210-19.

US EPA SW-846, Update III; Test Methods for Evaluating Solid Waste, Method 3545; Fed. Reg. Vol. 62, 114, 32451 US GPO, Washington, DC, USA, June 13, 1997.

7 Microwave-assisted solvent extraction in organic analysis

John R. Dean, Lisa Fitzpatrick and Carolyn Heslop

7.1 Introduction

Whether simply heating up a meal or preparing samples for analysis via a multistep process, microwave technology has both domestic and environmental importance. It can reduce the processing time of many extractions and, therefore, has economic value for laboratories worldwide. The way in which microwaves enhance extraction is not fully understood. The main factors to consider include improved transport properties of molecules, molecular agitation, the heating of solvents above their boiling points and, in some cases, product selectivity.

Any extraction procedure must comply with strict criteria. These include the ability to remove analytes quantitatively and repeatably, whilst using solvents that are not detrimental to the environment. Industry also requires methods to be cost-effective, robust and easy to use, with the potential for automation. Microwave-assisted extraction (MAE) is a viable alternative to other procedures, such as Soxhlet and soxtec extraction. Soxhlet extraction, whilst having the ability to produce quantitative data, suffers from several disadvantages, including the use of large quantities of organic solvents, especially chlorinated solvents, such as chloroform or dichloromethane (DCM), and the relatively slow speed of the extraction process. MAE has an advantage over this conventional approach in that it uses a drastically smaller volume of alternative solvents, such as acetone.

7.2 Microwave interaction with matter

Microwaves are high frequency electromagnetic radiation with a typical wavelength of 1 mm to 1 m. Many microwaves, both industrial and domestic, operate at a wavelength of around 12.2 cm (or a frequency of 2.45 GHz) to prevent interference with radio transmissions (Zlotorzynski, 1995). Microwaves are split into two parts, the electric field and the magnetic field component. These are perpendicular to each other and the direction of propagation (travel) and vary sinusoidally. Microwaves are comparable to light in their characteristics. They are said to have particulate character as well as acting like waves. The 'particles' of microwave energy are known as photons. These photons are absorbed by the molecule in the lower energy state (E_0) and the energy raises an electron to a higher energy level (E_1). Since electrons occupy definite energy levels, changes in these levels are discrete and, therefore, do not occur continuously. The energy is said to be quantised. Only charged particles are affected by the electric field component of the microwave. The Debye equation for the dielectric constant of a material determines the polarisability of the molecule. If the charged particles or polar molecules are free to move, this causes a current in the material. However, if they are bound strongly within the compound and, consequently, are not mobile within the material, a different effect occurs. The particles re-orientate themselves so that they are in-phase with the electric field. This is known as dielectric polarisation (Jacob and Boey, 1995).

Dielectric polarisation is split into four components, each one based upon the four different types of charged particles that are found in matter. These are: electrons, nuclei, permanent dipoles and charges at interfaces. The total dielectric polarisation of a material is the sum of all four components:

$$\alpha_1 = \alpha_e + \alpha_a + \alpha_d + \alpha_i \tag{7.1}$$

where, α_1 is the total dielectric polarisation, α_e is electronic polarisation (polarisation of electrons round the nuclei), α_a is the atomic polarisation (polarisation of the nuclei), α_d is the dipolar polarisation (polarisation of permanent dipoles in the material), and α_i is the interfacial polarisation (polarisation of charges at material interfaces).

The electric field of microwaves is in a state of flux, i.e. it is continually polarising and depolarising. These frequent changes in the electric field cause similar changes in the dielectric polarisation. Electronic and atomic polarisation and depolarisation occur more rapidly than the variation in the electric field. They have no effect on the heating of the material. Interfacial polarisation (also known as the Maxwell-Wagner effect) only has a significant effect on dielectric heating when charged particles are suspended in a non-conducting medium, e.g. a surfactant in an organic solvent, and are subjected to microwave radiation. The time period of oscillation of permanent dipoles is similar to that of the electric field of microwaves. The resulting polarisation lags behind the reversal of the electric field and causes heating in the substance. These phenomena are thought to be the main contributors to dielectric heating.

7.2.1 Choice of reagents

A substance that absorbs microwave energy strongly is called a sensitizer. The sensitizer preferentially absorbs the radiation and passes the energy on to other molecules. Polar molecules and ionic solutions (usually acids) will absorb microwave energy strongly in relation to nonpolar molecules. This is because they have a permanent dipole moment that will be affected by the microwaves. If extraction between nonpolar molecules is required, then the choice of solvent is the main factor to consider. If the solvent molecule is not sensitive enough to the radiation, then there will be no extraction. This is because the substance will not heat up.

7.2.2 Solvent effects

The choice of solvent for MAE is fundamental. The solvent must be able to absorb microwave radiation and pass it on in the form of heat to other molecules in the system. The following equation (Caddick, 1995) measures how well a certain solvent will pass on energy to others:

$$\varepsilon''/\varepsilon' = \tan \delta \tag{7.2}$$

where, δ is the dissipation factor, ε'' is the dielectric loss (a measure of the efficiency of conversion of microwave energy into heat energy), and ε' is the dielectric constant (a measure of the polarisibility of a molecule in an electric field).

Polar solvents, such as water, acetone and methanol, all absorb microwaves readily and are heated up when subjected to microwave radiation. Nonpolar solvents, such as hexane and toluene, do not heat up when they are subjected to microwave irradiation. In order to control the extraction process, mixtures of polar and nonpolar solvents can be used, e.g. acetone : hexane (1:1 v/v). Since the extraction process typically takes place in a closed vessel, the solvent chosen can be heated well above its normal boiling point. This will reduce the time required for the extraction process. Table 7.1 lists some of the solvents most commonly used in MAE (Hasty and Revesz, 1995).

Table 7.1 Solvents commonly used in microwave-assisted extraction (MAE)

Solvent	Dielectric constant	Boiling point °C	Closed-vessel temperature °C
Hexane	1.89	68.7	–
Methanol	32.63	64.7	151
Acetone	20.7	56.2	164
Acetonitrile	37.5	81.6	194

7.3 MAE systems

Two types of microwave heating systems are commercially available: an open-focused and a closed vessel system. The Appendix contains a list of suppliers of microwave systems. In the open-style system, individual sample vessels are heated sequentially. A typical commercial system is the Soxwave from Prolabo Ltd, France, which operates at percentage power increments from 0 to 100%, corresponding to a maximum of 300 W. A schematic diagram of an open-focused microwave system is presented in Figure 7.1. These power increments can be operated in stages and for various time intervals (up to 9 h). The sample and solvent are introduced into a glass container, which has the appearance of a large boiling/test tube and is fitted with either an air or a water condenser to prevent loss of volatiles and of the solvent. The sample container is placed within a protective glass sheath. The organic solvent is then heated, by means of microwave energy, and refluxed through the sample.

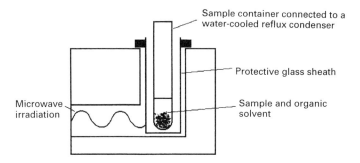

Figure 7.1 Open-focused microwave-assisted extraction system.

A common commercial closed system is the MES-1000 Microwave Solvent Extraction System, as supplied by CEM Corp., USA (Figure 7.2). This system allows up to 12 extraction vessels to be irradiated simultaneously, in 1% increments, up to 950 W of microwave energy at 100% power. The closed system has safety and important experimental features incorporated within its design, most notably an audible solvent alarm for the detection of any unexpected release of flammable and toxic organic solvent and also an ability to monitor both pressure and temperature *in situ* (within a single extraction vessel). The pressure is measured using a water manometer that allows readings of up to 200 psi to be made. The temperature probe, a fibreoptic with a phosphorus sensor (Papoutsis, 1984), allows extraction temperatures to be selected from 20 to 200°C in 1°C increments. The extraction conditions can be

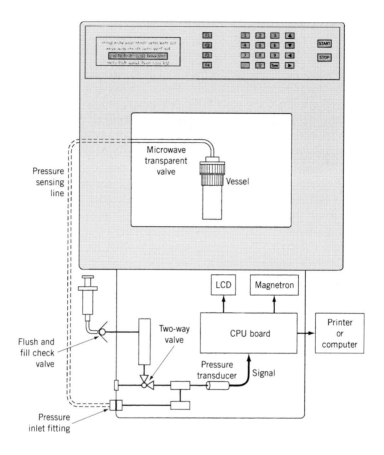

Figure 7.2 Microwave-assisted extraction: temperature and pressure control. (Courtesy of CEM Ltd, USA).

varied according to the percentage power input or by measuring the temperature and pressure within the single extraction vessel. The samples are placed into lined vessels (approximately 100 ml) constructed of polyetherimide (bodies and caps) (Figure 7.3). Inside each vessel is an inner liner and cover, constructed of Teflon perfluoroalkoxy (PFA), with which the sample comes into contact. Each extraction vessel contains a rupture membrane that is designed to fail at elevated pressure (200 psi). Each of the extraction vessels is located in a carousel, situated within the Teflon-lined microwave cavity, which rotates through 180° during microwave operation. In the centre of the carousel and connected to each extraction vessel is an expansion chamber, which acts to contain

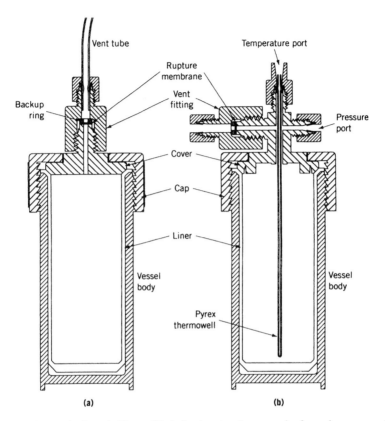

Figure 7.3 a) standard and b) modified lined extraction vessels for microwave-assisted extraction. (Courtesy of CEM Ltd, USA).

escaping vapours in the event of membrane rupture failure. In addition, a solvent detector system, located in the continuously operated air exhaust, will automatically turn-off the magnetron if solvent vapours from a leaking extraction vessel are detected within the microwave cavity. The exhaust fan will continue to operate in the event of a solvent escape. A schematic diagram of the safety features of the MAE system is presented in Figure 7.4.

7.4 Heating methods

The reduction in extraction time when using a microwave can possibly be attributed to the difference in heating methods employed by the

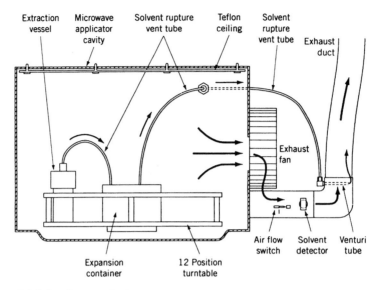

Figure 7.4 Safety features of microwave-assisted extraction. (Courtesy of CEM Ltd, USA).

microwave and conventional heating. Heating profiles (Figure 7.5) for water show that liquid heated in a microwave reaches its boiling point much more rapidly than under conventional heating methods. In conventional heating, e.g. with a Bunsen burner or isomantle, a finite period of time is required to heat the vessel before the heat is transferred to the solution. Thermal gradients are set up in the liquid due to convection currents. This means that only a small fraction of the liquid is at the required temperature.

Microwaves heat the solution directly, without heating the vessel, so that temperature gradients are kept to a minimum. Therefore, the rate of heating using microwave radiation is faster than in conventional methods and there is no loss of energy due to unnecessary heating of the vessel. Localised superheating can also occur (Kingston and Jassie, 1988).

7.5 Application of MAE for environmental analysis

The use of MAE for environmental analysis is a continuously expanding area of research at the present time. As a consequence, it is difficult to provide a totally up-to-date study. In order to compensate for this limitation, selected areas of environmental interest for which MAE has

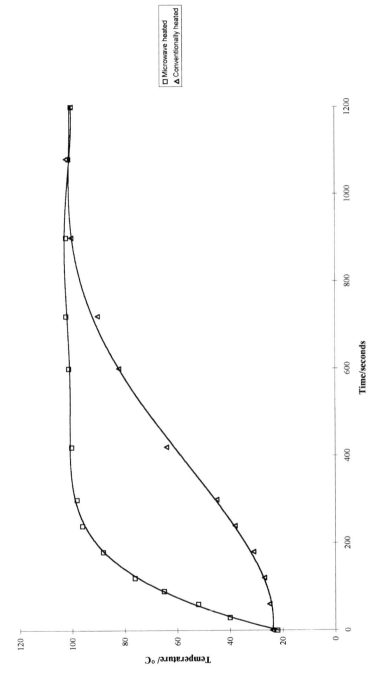

Figure 7.5 Heating profiles of deionised water. Comparison of microwave (–□–) and conventional (–△–) heating.

been utilised are highlighted in the text, with summaries provided in Tables 7.2 and 7.3.

7.5.1 Polycyclic aromatic hydrocarbons (PAHs)

As part of an on-going programme addressing new sample preparation techniques, the US Environmental Protection Agency (EPA) has evaluated MAE for the extraction of 17 PAHs and a few base/neutral/acidic compounds from spiked soil samples and certified reference materials (CRMs) (Lopez-Avila et al., 1994). The main MAE operating variables investigated were: temperature (80, 115 and 145°C) and extraction time (5, 10 and 20 min) using a 1+1 solvent mixture of hexane:acetone (30 ml). Six CRMs were evaluated: four sediments, HS-3, HS-4 and HS-5 available from the National Research Council of Canada (NRCC) and a National Institute of Standards and Testing (NIST) standard reference material (SRM) 1941; and two certified soils, SRS 103-100 and ERA, lot no. 321. The MAE recoveries were 70% at 80°C, 75% at 115°C and 75% at 145°C. As the recoveries at 115 and 145°C were identical, a temperature of 115°C for 10 min was used in subsequent work. Further work from the same group (Lopez-Avila et al., 1995a) using the same MAE conditions was carried out on 187 compounds, which included PAHs from spiked soils that were either extracted immediately or aged for varying lengths of time (24 h, 14 days or 21 days). It was found that ageing of the samples lowered the extraction efficiency.

The choice of solvent is an important criterion in the utilisation of any extraction technique. However, it is particularly important in MAE where a polar solvent is essential for microwave interaction. A commonly used solvent system for MAE is a 1+1 mixture of hexane and acetone. Barnabas and co-workers (1995) investigated the effect of varying this ratio on the recovery of PAHs from a native contaminated soil. The combinations were varied from 80:20 hexane:acetone to 0:100 hexane:acetone in steps of 10. It was found that as the solvent mixture became more polar, i.e. a hexane:acetone ratio of 10:90 or higher, it was possible to recover higher concentrations of PAHs from the native soil (Table 7.4). To ascertain the repeatability of the extraction, multiple extractions were performed using acetone. The results obtained gave a total recovery of 422.9 mg/kg, based on the sum of 16 individual PAHs with an average relative standard deviation (RSD) of 2.4% (n=6). It was concluded that the most effective solvent for the extraction of PAHs from natural soil was 100% acetone. To evaluate the three main operating conditions of MAE, i.e. temperature, extraction time and solvent volume, a chemometrics approach based on a central composite design was used, requiring 20 experiments. The operating conditions were varied according to the following: temperature, 40–120°C; extraction time, 5–20 min; and solvent volume, 30 and 50 ml. It was concluded that MAE using 100% acetone was insensitive to the variables investigated and that excellent recoveries could be obtained from a native contaminated soil.

A further approach, using a commercial MAE system, to optimise the extraction of PAHs from two marine sediment CRMs was reported using a mixed level orthogonal design (Chee *et al.*, 1996a). The main MAE variables investigated were: types of extraction solvent; extraction temperature (115 and 135°C); extraction time (5 and 15 min); and volume of extraction solvent (30 and 45 ml). The extraction solvents considered were: DCM, acetone-hexane (1 + 1), acetone-petroleum ether (1 + 1) and methanol-toluene (9 + 1). The optimum MAE conditions for the extraction of 16 PAHs from marine sediment (HS-4 and HS-6, available from the NRCC) were determined to be as follows: 30 ml of hexane:acetone (1 + 1); temperature, 115°C; and an extraction time of 5 min. The optimised approach was compared with a 16 h Soxhlet extraction using 300 ml of DCM as the solvent. Comparable results were obtained by both extraction methods. In addition, the speed of extraction (12 samples in one run of less than 30 min) and the lower solvent consumption (30 ml per sample) indicate the superiority of MAE.

7.5.2 Pesticides

Sample ageing is an important factor when evaluating any new extraction technique, as it allows some degree of analyte-matrix interaction to occur and, therefore, provides a more realistic environmental situation. In this context, Onuska and Terry (1993) aged a slurry spiked air-dried sediment for at least one month prior to MAE extraction of organochlorine pesticides. In contrast to other workers, Environment Canada have patented the extraction process (Pare *et al.*, 1991), and refer to the technique as the Microwave Assisted Process (MAP™). The workers highlight the dependence of extraction efficiency on the following: extraction solvent (a 1:1 mixture of iso-octane:acetonitrile being preferred to individual solvents); the sample requires a minimum water content for maximum recovery to be effected (a 15% water level was determined to be optimum); and extraction time (> 3 min). The results for the extraction of 15 organochlorine pesticides (OCPs) from a spiked sediment sample (at spiking levels between 50 and 250 μg/kg) were good; the minimum recovery was 74% (3% RSD) for p,p'-dichlorodiphenylchloroethane (p,p'-DDE), while the maximum recovery was 95.3% (3.9% RSD) for methoxychlor (n = 5).

Lopez-Avila and co-workers (1995b) reported results for the recovery of 20 OCPs from spiked soils (spike level 50 ng/g). The soil types used

Table 7.2 Microwave-assisted extraction (MAE) of organic contaminants from environmental and related samples

Analyte	Matrix	MAE sample preparation conditions	Comments	Reference
PAHs (n = 17) and other base/neutral compounds	Soil and sediments	Extraction of samples (5 g) using hexane:acetone (1:1) at temperatures of 80–145°C and times of 5–20 min at 50% power.	Room temperature extraction gave average recoveries of 52%, whereas MAE had recoveries for 17 PAHs of 70, 75 and 75% at 80, 115 and 145°C, respectively. Performance varied with the analyte and matrix.	Lopez-Avila et al. (1994)
PAHs (n = 16)	Soil	Native soil samples (2 g) extracted using acetone, DCM or hexane:acetone.	MAE with acetone at 120°C for 20 min gave the best results. A central composite design was used to elucidate the optimum operating parameters. Comparison with Soxhlet extraction.	Barnabas et al. (1995)
PAHs (n = 16)	Soil	Native soil samples (2 g) extracted using acetone or DCM at 120°C for 20 min.	Comparison of MAE with Soxhlet (10 g sample, mixed with anhydrous sodium sulphate and extracted with DCM for 6 h) and SFE (1 g sample, extracted for 1 h at a pressure of 200 kg/cm^2, temperature 70°C with 20% methanol-modified CO_2). Total recoveries, sum of 16 individual PAHs, were 297.4, 422.9 and 457 mg · kg^{-1} for Soxhlet, MAE and SFE,	Dean et al. (1995)

Analyte	Matrix	Conditions	Comments	Reference
PAHs (n = 15)	Polyurethane foam	Spiked samples (7.6 cm × 6 cm diameter) were extracted after 1, 6 and 21 days using cyclohexane. Spike levels ranged 100–2000 ng/ml.	Comparison with Soxhlet extraction (12 h extraction) and a mechanical plunging extraction approach (for 30 min). All extracts were concentrated to 1 ml prior to GC analysis. Microwave results were higher than the mechanical plunging approach for low molecular weight PAHs but similar to Soxhlet extraction. Low molecular weight PAHs exhibited low recoveries, probably due to their volatility. Typical %RSDs ranged 4.9–27.4, 2.4–5.4 and 2.2–30 for Soxhlet (n = 5), MAE (n = 6) and SFE (n = 7), respectively.	Lao et al. (1996)
PAHs (n = 16)	Sediment	Sample (5 g) extraction optimized using a mixed-level orthogonal array design procedure; solvent, temperature, duration and volume of solvent evaluated.	Results compared with Soxhlet extraction; higher extraction efficiency obtained using MAE. Recoveries from two certified reference samples (HS-4 and HS-6) were > 73.3%.	Chee et al. (1996a)

Table 7.2 (Continued)

Analyte	Matrix	MAE sample preparation conditions	Comments	Reference
PAHs, aliphatic hydrocarbons and pesticides	Lyophilised marine sediment	Samples (2–10 g) extracted with toluene and 1 ml water at 660 W power for 6 min.	Extracts dried with sodium sulphate, evaporated to dryness and residue dissolved in 2 ml hexane. Copper wires were added to remove sulphur, then the solution was cleaned-up using an activated-Florisil column. Elution with 5.5 ml hexane for aliphatic hydrocarbons and 30 ml DCM/hexane for PAHs and pesticides. Analysis by GC. Recoveries ranged from 97–102%	Pastor et al. (1997)
Total petroleum hydrocarbons	Soil	Samples (5 g) were extracted with 30 ml acetone:hexane (1:1) at 150°C for 5–15 min.	Comparable results to Soxhlet extraction; greater recoveries obtained by MAE than by sonication.	Hasty and Revesz (1995)
PCBs	Soil	Samples (5 g) extracted using hexane:acetone at 115°C for 10 min at 500 W power. Use of Weflon components for nonpolar solvent systems.	Extracts analysed by GC-MS after optional clean-up on AgNO$_3$-loaded silica gel 60 overlaid with alumina. Recoveries were good and the method was faster than Soxhlet extraction.	Kopp and Lautenschlaeger (1996)

Atrazine and principal degradates	Soil	Samples extracted twice with organic-free water at 95–98°C and then three times with 0.35 N HCl. Acid extracts were adjusted to pH 7, combined with the water extracts, centrifuged, filtered and concentrated using solid phase extraction.	Recoveries ranged 85–115% for deisopropylatrazine and de-ethylatrazine and 50–65% for atrazine and terbutylatrazine, with typical RSDs of 30–40%.	Steinheimer (1993)
Triazine herbicides	Soil	Samples (10 g) extracted with 40 ml of DCM:methanol (9:1) at 115°C for 20 min at full power and 100 psi pressure	Extract dried with anhydrous sodium sulphate and concentrated to 2 ml using a Kuderna-Danish apparatus. The extract was then taken to dryness under a stream of N_2 and the residue reconstituted with 5 ml of methyl t-butylether prior to GC analysis. Recoveries ranged 89–103% at a spiking level of 200 ng/kg with RSDs of 2.1–5.3%	Molines et al. (1996)
Imidazolinone herbicides	Soil	Spiked samples (20 g) extracted with 0.1 M NH_4OAc/NH_4OH at 125°C for 3 min.	Average recovery of 92% with an RSD of 13%, based on spike levels of 1–50 ng/ml.	Stout et al. (1996)

Table 7.2 (Continued)

Analyte	Matrix	MAE sample preparation conditions	Comments	Reference
Hexaconazole	Soil	Sample (5 g) extracted with 30 ml acetone at 115°C for 15 min at 1000 W.	Weathered soil samples extracted by Soxhlet, MAE, SFE and ASE. Some matrix dependency was noted for MAE from the sandy loam soil (organic matter 5.7%).	Frost et al. (1997)
4-nonylphenol	Sediment	Sample (5 g) extracted with 30 ml of DCM at 100°C for 5 min.	Comparison with Soxhlet extraction (16 h using 300 ml of DCM). Optimisation of process using a two-level orthogonal array design. Recoveries were >80%.	Chee et al. (1996b)
Phenol and methylphenol	Soil	Aged (25 days) samples (1–5 g) extracted using 10–50 ml of acetone:hexane at 130°C for 15 min.	MAE operating variables optimized using a factorial design approach. Recoveries ranging 94.5–104.4%, with RSDs 6.1–9.9% (n = 5), compared favourably with sonication recoveries (45.3–58.7%).	Llompart et al. (1997a)
Phenol and methylphenol	Soil	Aged (20 days) samples (0.5–5 g) extracted and derivatised in situ using pyridine, acetic anhydride and hexane (9 ml) at 130°C for 30 min.	MAE operating variables optimized using a factorial design approach. Recoveries from freshly spiked samples ranged 90.5–112.1%, with RSDs 4.0–21.4%, irrespective of spike level (40 or 400 ng/g).	Llompart et al. (1997b)

187 organic compounds and four aroclors	Soil	Extraction with hexane:acetone (1:1) at 115°C for 10 min.	Recoveries of 79 out of 95 compounds from freshly spiked top soil were 80–120%; 14 recoveries were <80% (5 of these <20%); and one recovery was >120%. Recoveries from aged samples were inferior, except for six compounds. For 38 out of 45 organochlorine compounds and 34 out of 47 organophosphorus pesticides recoveries were 80–120%. Recoveries of aroclors were 75–157%.	Lopez-Avila et al. (1995a)
94 organic pollutants (listed in EPA Method 8250)	Soil and sediments	Sample (10 g) extracted with hexane:acetone (1:1) at 115°C for 10 min. After cooling to room temperature, the supernatant was mixed with 3–5 ml of hexane:acetone and concentrated to 5 ml by evaporation under N_2 and centrifugation for 10 min at 2300 rpm. Extract concentrated to 1 ml prior to GC analysis.	Comparison with Soxhlet (DCM:acetone, 1:1), sonication and SFE (10% methanol-modified CO_2). MAE gave recoveries for 51 compounds of >80%; 33 compounds, 50–79%; eight compounds 20–49%; and two compounds <19%.	Lopez-Avila et al. (1996)

Table 7.2 (Continued)

Analyte	Matrix	MAE sample preparation conditions	Comments	Reference
Volatile organic compounds and BTEX compounds	Soil	Microwave-assisted process gas phase extraction. Samples (1 g) immersed in 15 ml of water in a headspace vial irradiated for 48 s at 50% power.	A detection limit of 0.020 μg/g was obtained, with an RSD of 10%. Linear range was 0.2–20 ppm.	Pare et al. (1996/97)
Salinomycin	Animal feedstuffs	Extraction of finished chicken feedstuff (0.6 g) with 15 ml ethanol and 2 ml of propan-2-ol followed by irradiation for 8 s at 800 W power. Organic layer decanted off and process repeated twice more. Extracts combined.	Extracts concentrated to dryness in vacuo and reconstituted in HPLC mobile phase. Quantitative recoveries obtained.	Akhtar and Croteau (1996)
Phthalate esters	Soil and sediment	Samples (5 g) extracted with hexane:acetone (1:1) at 115°C for 10 min.	After centrifugation and concentration to 1 ml, extracts analysed by GC. Recoveries ranged 65.5–89.5%, and this compared favourably with those obtained by Soxhlet (65.5–89.5%) and sonication (64.6–88.6%).	Chee et al. (1996c)

Abbreviations: PAH, polycyclic aromatic hydrocarbon; DCM, dichloromethane; GC, gas chromatography; PCB, polychlorinated biphenyl; SFE, supercritical fluid extraction; ASE, accelerated solvent extraction; EPA, Environmental Protection Agency; BTEX, benzene, toluene, ethylbenzene and xylene; HPLC, high performance liquid chromatography; RSD, relative standard deviation; MS, mass spectroscopy.

Table 7.3 Microwave-assisted extraction (MAE) of organic contaminants from aqueous samples

Analyte	Matrix	MAE sample preparation conditions	Comments	Reference
Volatile organic compounds	Water	Sample extracted using microwave-assisted process gas-phase extraction with headspace samplings. A microwave power of 500 W was used, with an irradiation and spinning cycle of 15 s irradiation, 5 s spinning, 15 s irradiation and 5 s spinning. Headspace sampling was carried out at 90°C.	Direct transfer of liberated volatiles to GC. Calibration graphs were linear for 10–5000 ppb, with typical RSDs of 1.4%.	Pare et al. (1995)
PCBs	Water	Samples (500 ml) were extracted with 50 ml of 2,2,4-trimethylpentane for 6×2 min at 65–75°C and full power.	Combined extracts were dried with sodium sulphate and concentrated to 3 ml. Column clean-up using silica gel was also performed. Recoveries from distilled water ranged 64.7–85.5% for standard PCB congeners.	Onuska and Terry (1995a)
Chlorinated benzenes	Water	Surface water was spiked with an internal standard and NaCl added. A helium purge line was introduced into the extraction vessel. Purged analytes were collected in ice-cold hexane. Samples were irradiated for 7 min at full power.	Extracts cleaned-up with a Florisil column and eluted with 20% toluene in iso-octane prior to GC analysis. Detection limits in pg/l obtained.	Onuska and Terry (1995b)

Table 7.3 (Continued)

Analyte	Matrix	MAE sample preparation conditions	Comments	Reference
PAHs, PCBs, phthalates and pesticides	Water	Membrane (Empore) solid phase extraction using MAE; microwave elution with 20 ml acetone for 5 min at 100°C	Analysis by GC or HPLC. Recoveries for 8 model compounds (anthracene, benzo[a]pyrene, dimethylphthalate, gamma-HCH, p,p′-DDT and PCB1260) added to water were 70–86%, with typical RSDs of <9.2%. Recoveries were improved in the presence of NaCl and decreased in the presence of humic acid.	Chee et al. (1996d)
Phenols	Water	Membrane (Empore) solid phase extraction using MAE; microwave elution with 10 ml methanol for 5 min at 100°C.	Comparison with conventional solid phase extraction on a C18 cartridge. Extracts evaporated to 1 ml under N_2 prior to HPLC analysis. Recoveries from spiked water, tap water and seawater were 86.5, 81.8 and 81.5%, respectively, for the Empore membranes. Typical RSDs ranged 7–9%.	Chee et al. (1997)

Abbreviations: GC, gas chromatography; PCB, polychlorinated biphenyl; PAH, polycyclic aromatic hydrocarbon; HPLC, high performance liquid chromatography; DDT, dichlorodiphenyl trichloroethane; gamma-HCH, gamma-hexachlorocyclohexane; p,p′-DDT, p,p′-dichlorodiphenyldichloroethane; RSD, relative standard deviation.

Table 7.4 Results of microwave-assisted extraction (MAE) using various compositions of acetone/hexane to extract soil (all extractions performed in duplicate)

Compound	Composition hexane/acetone							
	80/20	70/30	60/40	50/50	40/60	30/70	20/80	10/90	0/100
Naphthalene	6.7	8.5	9.1	8.4	9.5	11.2	11.2	13.7	13.5
Acenaphthylene	3.4	4.5	3.5	3.5	4.7	4.4	4.2	5.2	5.0
Acenaphthene	8.9	10.4	10.0	9.9	11.3	14.2	14.2	16.6	16.7
Fluorene	10.6	12.5	10.1	10.8	15.6	15.0	15.0	17.6	17.5
Phenanthrene	61.2	54.3	56.7	57.4	57.6	72.7	73.7	85.5	86.6
Anthracene	15.7	17.6	18.6	18.3	20.7	22.9	22.8	26.4	26.4
Fluoranthene	51.5	61.1	59.6	58.8	61.0	75.2	74.8	87.6	89.2
Pyrene	39.5	49.3	49.3	48.1	51.2	61.5	61.0	69.2	70.2
Benz(a)anthracene	21.9	21.0	21.4	20.2	20.6	26.6	26.4	24.6	25.3
Chrysene	25.0	21.0	19.9	19.0	20.2	26.9	26.5	24.8	25.4
Benzo(b)fluoranthene	12.1	17.2	18.3	17.4	19.3	14.5	15.0	18.9	13.7
Benzo(k)fluoranthene	9.6	13.7	12.8	13.2	13.8	15.4	15.6	8.8	10.4
Benzo(a)pyrene	12.5	19.9	19.5	18.6	20.0	26.0	26.0	24.1	24.6
Indeno (1,2,3-cd)pyrene	5.8	9.3	8.7	8.6	9.5	17.0	17.0	15.8	15.7
Dibenz(a,h)anthracene	2.5	3.5	3.2	3.0	3.7	4.2	4.2	3.8	3.8
Benzo(ghi)perylene	6.9	9.4	8.5	8.1	9.1	14.7	14.5	13.6	13.9
Total	293.4	332.6	328.9	323.1	347.4	421.9	421.7	455.7	457.6

Concentrations are in mg/kg.

were varied, i.e. a clay soil, topsoil, sand, organic compost and topsoil with 5% humic acid. The following MAE conditions were used: 30 ml of a 1:1 acetone:hexane mixture; temperature, 115 °C; and an extraction time of 10 min at 100% power. The cleanest extracts and highest recoveries (mean recovery 83.4% with an RSD of 10.6%) were obtained from the sand matrix. Recoveries from the other matrices were only slightly different. The mean recovery from the clay soil was 85.1% with an RSD of 20%, while from the organic compost the mean recovery was 88.8% with an RSD of 30.0%. A significantly poorer recovery was obtained both from the top soil with 5% humic acid added (mean recovery 71.9% with an RSD of 24.9%) and the topsoil only (mean recovery 72.9% with an RSD of 20.4%).

In order for MAE to be useful as an extraction technique, it requires the use of organic solvents with a permanent dipole moment. In an interesting departure from this approach, Hummert and co-workers (1996) have utilised a microwave transformer (Weflon®, an inert material based on carbon-containing Teflon) to transfer energy to the sample in the presence of a nonpolar solvent, hexane. This approach was applied for the extraction of OCPs from seal blubber and pork fat. The MAE system was operated in extraction cycles of short duration, 30 s at full power followed by 5 min cooling. In order to assess the accuracy of this approach, the blubber of a grey seal stranded on Rugen, Baltic Sea, Germany was used. This seal blubber was initially melted and subjected to matrix clean-up with deactivated silica prior to analysis. The results from this approach were compared to those obtained using MAE. It was experimentally determined that the best approach for MAE was to operate with seven extraction cycles. Average recoveries for the seven organochlorine compounds were $96.9 \pm 0.5\%$. Recoveries of organochlorine compounds from spiked pork fat ranged from 88.5 ± 1.3 to 98.6 ± 2.3, based on seven extraction cycles.

The extraction of the herbicide, atrazine, and its metabolites (deethylatrazine and deisopropylatrazine) from spiked agricultural soil was reported by Steinheimer (1993). Nashua soil: organic carbon content, 4.48%; cation exchange capacity, 36.1 mEq/100 g; pH, 6.7; sand, 42.5%; silt, 47.5%; and clay, 10.0%; and Treynor soil: organic carbon content, 1.81%; cation exchange capacity, 32.3 mEq/100 g; pH, 6.1; sand, 10.0%; silt, 77.5%; and clay, 12.5%, were used. Both soils were extracted using a microwave oven, operating at a temperature of 95–98°C, with organic-free water and 0.35 N HCl as the extraction solvent. Combined extracts, of approximate volume 125 ml, were adjusted to pH 7.0, centrifuged, filtered and preconcentrated using solid phase extraction (SPE) prior to analysis by high-performance liquid chromatography (HPLC). Soil type did not appear to influence the recovery of atrazine and its metabolites. Approximate recoveries from the Nashua soil (n = 4) were 62% for atrazine, 95% for deethylatrazine and 88% for deisopropylatrazine. Recoveries from the Treynor soil were approximately 58, 83 and 118% for atrazine, deethylatrazine and deisopropylatrazine, respectively. It was also noted that the microwave extraction resulted in cleaner extracts and simpler chromatograms than produced with the more traditional solid-liquid extraction methods using methanol-water and acetonitrile-water.

The extraction of a new class of low-use-rate, reduced environmental-risk herbicides, the imidazolinones, has been reported (Stout *et al*., 1996). As the level of application is low, it is required that the herbicides can be determined in soil at the low ppb level. The current methodology uses a 1 h extraction of 50 g of soil with 200–300 ml of 0.5 N NaOH, followed by extensive clean-up using precipitation and centrifugation. This is followed by partitioning with 200–300 ml of DCM. After concentration, the final extract is cleaned-up using SPE (a strong cation exchange column and a C18 column are both required). This extensive extraction, concentration and clean-up procedure allows a level of quantitation of 5 ppb using HPLC with UV detection. It is estimated that the preparation of six samples takes $\sim$12 h. Any improvements in this procedure are, therefore, advantageous. Using the test analyte, imizethapyr, MAE was applied to spiked soils in the concentration range 1–50 ppb. Optimum MAE conditions were found to be as follows: temperature, 125 °C; extraction time, 3 min; and an extraction solvent of 0.1 M NH_4OAc/NH_4OH at pH 10. These conditions gave acceptable recoveries (mean recovery $92 \pm 13\%$, n = 12).

7.5.3 Polychlorinated biphenyls

Polychlorinated biphenyls (PCBs) have been extracted from a range of CRMs (marine sediment and soils) using MAE. The following conditions were used: solvent, 30 ml of hexane:acetone (1 + 1); temperature 115°C; and extraction time, 10 min (Lopez-Avila *et al*., 1995a). The results, for a range of Aroclors (1254, 1260, 1016, 1248), were in good agreement with certified values (Table 7.5). Lopez-Avila and co-workers (1995c) have compared the method of detection (gas chromatography-electron capture detection [GC-ECD] or enzyme-linked immunosorbent assay [ELISA]) of the PCBs after MAE. The results are shown in Table 7.6. It was concluded that the rapidity of MAE coupled with ELISA allows a batch of 10 samples to be prepared and analysed in approximately 1 h.

7.5.4 Phenols

Optimisation of the extraction method can be effectively achieved using chemometrics. It is, therefore, not surprising to find that chemometrics has been utilised for MAE optimisation. Llompart and co-workers

Table 7.5 Average recoveries of polychlorinated biphenyls (PCBs) from soil by microwave-assisted extraction (MAE)

Aroclor type	Matrix	Certified value or spike level, mg/kg	Recovery[a] %	% RSD
1254	HS-1 marine sediment[b]	0.022	93.2[c]	8.1
1254	HS-2 marine sediment[b]	0.112	76.7	6.0
1260	ERA soil (lot No. 9801)[d]	394	89.9[c,e]	2.6
1016	Freshly spiked topsoil[f]	0.100	85.9[e]	5.6
1260	Freshly spiked topsoil[f]	0.100	82.5[e]	4.2
1016	Spiked topsoil aged for 24 h at 4°C[f]	0.100	92.8[e]	14
1260	Spiked topsoil aged for 24 h at 4°C[f]	0.100	88.6[e]	10
1248	Superfund site sample 1[g]	465	102	3.5
1248	Superfund site sample 2[g]	1.13	157	6.3
1248	Superfund site sample 3[g]	0.033	86.8	35
1248	Superfund site sample 4[g]	6.7	75.3	17

[a]number of determinations = 3; [b]certified by the National Research Council of Canada (NRCC); [c]number of determinations = 4; [d]certified by Environmental Research Association (ERA); [e]recoveries were corrected for losses during blowdown evaporation; [f]spiked at the Midwest Research Institute, California; [g]reported by independent laboratory that used Soxhlet extraction and GC/ECD. (Lopez-Avila et al., 1995a).
Abbreviations: RSD, relative standard deviation; GC, gas chromatography; ECD, electron capture detection.

Table 7.6 Comparison of microwave-assisted extraction (MAE) ELISA and MAE GC ECD for extraction of PCBs from reference samples

Compound	Matrix	Certified value mg/kg	MAE ELISA (n = 3)		MAE GC-ECD (n = 3)	
			Conc. mg/kg	% RSD	Conc. mg/kg	% RSD
Aroclor 1260	ERA soil (Lot No. 9801)	394000	433000	7.0	326000	2.0
Aroclor 1254	HS-1 marine sediment	21.8	5.9	14.0	20.0	8.0
Aroclor 1254	HS-2 marine sediment	112	53.0	14.0	81.0	13.0

Abbreviations: ELISA, enzyme-linked immunosorbent assay; GC, gas chromatography; ECD, electron capture detection. (From Lopez-Avila et al., 1995c).

(1997a) utilised a central composite design for the MAE of phenol and methylphenol isomers from spiked soils that had been aged for 25 days. The various extraction conditions were as follows: temperature, 70–130°C; proportion of acetone, 20–80% with hexane making up the remainder; and solvent volume, 15–50 ml; at a fixed extraction time of 10 min and 5 g of sample. It was concluded, at the 95% confidence interval, that temperature was significant for all four phenols studied, while the ratio of acetone to hexane and the solvent volume were found to be significant in some cases. Quantitative recoveries were obtained using 10 ml of acetone:hexane (80:20) at 130°C.

7.5.5 Phthalate esters

The effects of extraction solvent, solvent volume, temperature and extraction time have been evaluated for the removal of six phthalate esters, by MAE, from spiked marine sediment (Chee *et al.*, 1996c). The phthalates investigated were: dimethyl phthalate, diethyl phthalate, diallyl phthalate, dibutyl phthalate, benzyl-n-butyl phthalate and bis (2-ethylhexyl)phthalate. The MAE operating variables investigated were as follows: extraction solvent (DCM, acetone/hexane and acetone/petroleum ether); solvent volume (25, 30 and 35 ml); temperature (80, 115 and 145°C); and extraction time (5, 10 and 15 min). In terms of extraction solvent, no statistical difference was observed between a 1:1 acetone:hexane mixture and DCM; however, it was preferable to use the solvent mixture. The solvent mixture was chosen as it required no solvent exchange prior to analysis. Additionally, DCM is not a suitable solvent for GC when using an electron capture detector. Optimum MAE conditions were identified as follows: 30 ml of a 1:1 solvent mixture (hexane:acetone); and temperature of 115°C for 10 min. This approach was then applied to marine sediment located in the Tuas/Jurong industrial area of Singapore. It was found that dibutyl phthalate and bis(2-ethylhexyl) phthalate were present in the ranges 0.68–1.60 and 0.16–2.79 mg/kg, respectively.

7.6 Microwave-assisted solid phase extraction

Microwave assisted extraction has been combined with SPE for the extraction of pollutants from aqueous samples (Chee *et al.*, 1996d). Initially, the analytes are retained on a C18 membrane disk (Empore), rolled up and transferred into the closed polytetra fluoroethane (PTFE)-lined vessels. Elution is achieved by placing organic solvent into the extraction vessel and applying microwave energy. This approach has been applied for the extraction of OCPs, PCBs, PAHs, phthalate esters, organophosphorus pesticides (OPPs), fungicides, herbicides and insecticides from aqueous samples (Chee *et al.*, 1996d). This alternative approach, microwave-assisted elution has been investigated in terms of the elution solvent (acetone and DCM) and microwave operating parameters. The following microwave parameters were investigated: temperature (80, 100 and 120°C) and extraction time (1, 3, 5 and 10 min) at 50% power. It was concluded that the optimum conditions were: solvent, acetone; temperature, 100°C; and, extraction time, 7 min. Favourable results were obtained, as compared to traditional liquid-liquid extraction, for PAHs and phthalates (spiked at 2 μg/l), OCPs and PCBs (spiked at 0.1–0.2 μg/l for OCPs and 0.5 μg/l for PCBs) and OPPs, fungicides, insecticides and herbicides (spiked at 1–2 μg/l) in either reagent water or sea water.

7.7 Gas-phase microwave-assisted extraction

The liberation of volatile organic compounds (VOCs) from water samples by microwave energy has been described by Pare *et al.* (1995). In this process, the aqueous sample is heated, via a microwave oven, and VOCs are vaporized into the headspace of the sample. By utilising a conventional headspace sampler the VOCs are introduced directly into a GC with flame ionisation detection (FID). This approach was compared with a conventional 30 min static headspace sampling apparatus and the results obtained were favourable. It was noted, however, that the microwave approach gave higher detector responses, with better precision and a shorter time scale than the conventional approach.

7.8 Future prospects for MAE

Microwave-assisted extraction has been extensively applied to a range of sample types (soils, sediments, water) for the environmental analysis of pollutants. The most common approach is the use of a pressurised microwave system. Frequently, the investigators have evaluated the dependence of operating parameters on analyte recovery. It is, therefore, possible to suggest some recommendations for the utilisation of pressurised MAE in the extraction of pollutants from solid matrices e.g. soils: temperature, $>115°C$ but $<145°C$; pressure, operating at <200 psi; microwave power, 100%; extraction time (time at parameter), >5 min but no need to extend beyond 20 min (the longest time is recommended if 12 vessels are to be extracted simultaneously); extraction solvent volume, 30–45 ml per 2–5 g of sample; and extraction solvent, hexane-acetone (1:1 v/v) has been most commonly used but other solvents also appear to be satisfactory (acetone and DCM).

These parameters are based on the use of a microwave system capable of delivering a minimum of 900 W of power. Ideally, the oven cavity should be equipped to allow ventilation of the cavity in the event of an organic vapour release. An additional safety feature is the inclusion of a solvent sensor that automatically shuts off the microwave source in the event of an organic solvent leakage, thus minimising the risk of fire. Extraction vessels should be mounted on a carousel arrangement that rotates through 360°, allowing equal dissipation of the microwave energy to each sample vessel.

References

Akhtar, M.H. and Croteau, L.G. (1996) Extraction of salinomycin from finished layers ration by microwave solvent extraction followed by liquid chromatography. *Analyst*, **121** 803-806.

Barnabas, I.J., Dean, J.R., Fowlis, I.A. and Owen, S.P. (1995) Extraction of polycyclic aromatic hydrocarbons from highly contaminated soils using microwave energy. *Analyst*, **120** 1897-904.

Caddick, S. (1995) Microwave-assisted organic reactions. *Tetrahedron*, **51** (38) 10403-32.

Chee, K.K., Wong, M.K. and Lee, H.K. (1996a) Optimization of microwave-assisted solvent extraction of polycyclic aromatic hydrocarbons in marine sediments using a microwave extraction system with high-performance liquid chromatography-fluorescence detection and gas chromatography-mass spectrometry. *J. Chromatogr.*, **723** 259-71.

Chee, K.K., Wong, M.K. and Lee, H.K. (1996b) Optimization of sample preparation techniques for the determination of 4-nonylphenol in water and sediment. *J. Liq. Chromatogr.*, **19** 259-75.

Chee, K.K., Wong, M.K. and Lee, H.K. (1996c) Microwave extraction of phthalate esters from marine sediment and soil. *Chromatographia*, **42** 378-84.

Chee, K.K., Wong, M.K. and Lee, H.K. (1996d) Microwave-assisted solvent elution technique for the extraction of organic pollutants in water. *Anal. Chim. Acta*, **330** 217-27.

Chee, K.K., Wong, M.K. and Lee, H.K. (1997) Membrane solid-phase extraction with closed vessel microwave elution for the determination of phenolic compounds in aqueous matrices. *Mikrochimica Acta*, **126** 97-104.

Dean, J.R., Barnabas, I.J. and Fowlis, I.A. (1995) Extraction of polycyclic aromatic hydrocarbons from highly contaminated soils: comparison between Soxhlet, microwave and supercritical fluid extraction techniques. *Anal. Proc.*, **32** 305-308.

Frost, S.P., Dean, J.R., Evans, K.P., Harradine, K., Cary, C. and Comber, M.H.I. (1997) Extraction of hexaconazole from weathered soils: a comparison between Soxhlet extraction, microwave-assisted extraction, supercritical fluid extraction and accelerated solvent extraction. *Analyst*, **122** 895-98.

Hasty, E. and Revesz, R. (1995) Total petroleum hydrocarbon determination by microwave solvent extraction. *American Laboratory*, **27** 66-73.

Hummert, K., Vetter, W. and Luckas, B. (1996) Fast and effective sample preparation for determination of organochlorine compounds in fatty tissue of marine mammals using microwave extraction. *Chromatographia*, **42** 300-304.

Jacob, J. and Boey, F. (1995) Thermal and non-thermal interaction of microwave radiation with materials. *J. Materials Science*, **30** 5321-27.

Kingston, H.M. and Jassie, L.B. (eds.) (1988) *Introduction to Microwave Sample Preparation*, American Chemical Society, Washington, DC, USA.

Kopp, G. and Lautenschlaeger, W. (1996) Microwave-assisted extraction. *LaborPraxis*, **20** 82-85.
Lao, R.C., Shu, Y.Y., Holmes, J. and Chiu, C. (1996) Environmental sample cleaning and extraction procedures by microwave-assisted process (MAP) technology. *Microchemical Journal*, **53** 99-108.
Llompart, M.P., Lorenzo, R.A., Cela, R. and Pare, J.R.J. (1997a) Optimization of a microwave-assisted extraction method for phenol and methylphenol isomers in soil samples using a central composite design. *Analyst*, **122** 133-37.
Llompart, M.P., Lorenzo, R.A., Cela, R., Pare, J.R.J., Belanger, J.M.R. and Li, K. (1997b) Phenol and methylphenol isomers determination in soils by *in situ* microwave-assisted extraction and derivatisation. *J. Chromatogr.*, **757** 153-64.
Lopez-Avila, V., Young, R. and Beckert, W.F. (1994) Microwave-assisted extraction of organic compounds from standard reference soils and sediments. *Anal. Chem.*, **66** 1097-106.
Lopez-Avila, V., Young, R., Benedicto, J., Ho, P., Kim, R. and Beckert, W.F. (1995a) Extraction of organic pollutants from solid samples using microwave energy. *Anal. Chem.*, **67** 2096-102.
Lopez-Avila, V., Young, R., Kim, R. and Beckert, W.F. (1995b) Accelerated extraction of organic pollutants using microwave energy. *J. Chromatogr. Sci.*, **33** 481-84.
Lopez-Avila, V., Benedicto, J., Charan, C., Young, R. and Beckert, W.F. (1995c) Determination of PCBs in soils/sediments by microwave-assisted extraction and GC/ECD or ELISA. *Environ. Sci. Technol.*, **29** 2709-12.
Lopez-Avila, V., Young, R. and Teplitsky, N. (1996) Microwave-assisted extraction as an alternative to Soxhlet, sonication and supercritical fluid extraction. *J. AOAC Int.*, **79** 142-56.
Molines, O., Hogendoorn, E.A., Heusinkveld, H.A.G., vanHarten, D.C., vanZoonen, P. and Baumann, R.A. (1996) Microwave-assisted solvent extraction (MASE) for the efficient determination of triazines in soil samples with aged residues. *Chromatographia*, **43** 527-32.
Onuska, F.I. and Terry, K.A. (1993) Extraction of pesticides from sediments using a microwave technique. *Chromatographia*, **36** 191-94.
Onuska, F.L. and Terry K.A. (1995a) Microwave extraction in analytical chemistry of pollutants: polychlorinated biphenyls. *J. High Res. Chromatogr.*, **18** 417-21.
Onuska, F.L. and Terry K.A. (1995b) Microwave extraction in analytical chemistry of pollutants: chlorinated benzenes. *J. Microcol. Sep.*, **7** 319-26.
Papoutsis, D. (1984) Fibreoptic trends: keeping the heat on cancer. *Photonics Spectra*, **March** 53-59.
Pare, J.R.J., Sigouin, M. and Lapointe, J. (1991) US Patent 5,002,784, March 26, Environment Canada.
Pare, J.R.J., Belanger, J.M.R., Li, K. and Stafford, S.S. (1995) Microwave-assisted process (MAP): application to the headspace analysis of VOCs in water. *J. Microcol. Sep.*, **7** 37-40.
Pare, J.R.J., Belanger, J.M.R., Li, K., Llompart, M.P., Singhvi, R. and Turpin, R.D. (1996/97) Gas-phase extraction method using the microwave-assisted process (MAP) for the determination of aromatic contaminants in soil. *Spectroscopy*, **13** 89-98.
Pastor, A., Vasquez, E., Ciscar, R. and De la Guardia, M. (1997) Efficiency of the microwave-assisted extraction of hydrocarbons and pesticides from sediments. *Anal. Chim. Acta*, **344** 241-49.
Steinheimer, T.R. (1993) HPLC determination of atrazine and principal degradates in agricultural soils and associated surface and ground water. *J. Agric. Food Chem.*, **41** 588-95.
Stout, S.J., daCunha, A.R. and Allardice, D.G. (1996) Microwave-assisted extraction coupled with gas chromatography/electron capture negative chemical ionization mass spectrometry for the simplified determination of imidazolinone herbicides in soil at the ppb level. *Anal. Chem.*, **68** 653-58.
Zlotorzynski, A. (1995) The application of microwave radiation to analytical and environmental chemistry. *Crit. Rev. Anal. Chem.*, **25** (1) 43-76.

Appendix

Suppliers of commercial microwave extraction systems

CEM Corporation, 3100 Smith Farm Road, P.O. Box 200, Matthews, NC 28106-0200, USA.

Prolabo Corporation, 24 Magnolia Ct, Lawrenceville, NJ 08648, USA.

Milestone Corporation, 7289 Garden Road, Suite 219, Riviera Beach, FL 33404, USA.

Questron Corporation, P.O. Box 2387, Princeton, NJ 08543-2387, USA.

8 Biological/pharmaceutical applications
D. Stevenson, S. Miller and I.D. Wilson

8.1 Introduction

The analysis of trace organics in biological fluids may be carried out for a variety of important purposes. These commonly include measurements of drugs and their metabolites, endogenous compounds, industrial chemicals (or suitable metabolites) and markers of environmental exposure.

Analyses of numerous drugs and metabolites are necessary for pharmacokinetic studies when candidate drugs are under development. Initial studies look at the profile in animals, before progressing to human volunteers and eventually patients. Such studies typically determine the concentration of drugs and, if appropriate, metabolites after administration. Thus, the appropriate dose and the timing between the administration of particular drugs can be selected. During drug development, it is also necessary to identify the metabolites produced after the drug has been administered. These metabolites will generally be more polar than the parent drug and may require a different strategy for sample preparation. However, it is clearly preferable to carry out analysis of both drug and metabolites simultaneously whenever possible.

Drug (or, if more appropriate, metabolite) concentrations are also measured for pharmaceuticals once they are on the market. This might be to check compliance with medication or, in the case of drugs with a narrow therapeutic ratio, to ensure that the concentration is above the level needed for therapeutic benefit but not so high that toxic effects might be encountered. This therapeutic monitoring, as with pharmacokinetic analysis, is necessary because of interindividual differences in the absorption, distribution, metabolism and excretion of drugs. When carrying out therapeutic drug monitoring, it should be remembered that drugs may be used in combination rather than just alone. It may be desirable, therefore, to determine more than a single drug or its metabolites in the same sample.

In a rather different situation, drugs of abuse are determined as part of the control measures to limit their use, or in forensic cases to investigate the possibility of deliberate or accidental overdoses or poisoning. In these cases, concentrations will usually be higher than when measuring therapeutic levels. The measurement of endogenous compounds in biological fluids is also quite common. This might be carried out to assist in the diagnosis of disease or the monitoring of treatment, or to investigate the biological changes induced by treatment with drugs. The study of biomarkers of particular conditions is an area of growing interest.

Although the purpose is rather different, modern industrial hygiene also utilises trace organic analysis in biological fluids. Industrial hygiene is concerned with monitoring the risk to human health arising from occupational exposure to chemicals. For many chemicals, occupational exposure standards are based on current knowledge of the short- and long-term toxicity of particular chemicals. Most occupational exposure standards are based on exposure via inhalation, and are thus the result of air monitoring. However, a significant (and growing) number are based on biological monitoring. This requires determination of the compound or metabolite in biological fluid, typically urine or plasma. For biological monitoring to be of use, the toxicokinetics and metabolism of the compound must be well characterised. It must be possible to relate the concentration in biological fluid to the original dose, and also to establish a 'no effect' level. Once a safe level has been established, a monitoring strategy can be implemented to ensure compliance. Unlike air monitoring, biological monitoring provides data directly related to individuals. It is also a more suitable method when dermal rather than respiratory exposure is the major route.

A number of studies have been carried out to measure trace organics in biological fluids from non-occupationally-exposed individuals. This would typically arise from environmental exposure to chemicals, such as pesticides, polycyclic aromatic hydrocarbons (PAHs) and polychlorinated biphenyls (PCBs). This is almost identical to the biological monitoring carried out to evaluate workplace exposure, except that the concentrations are hopefully lower and the range of compounds encountered is much greater. The procedures and equipment for biological monitoring, both occupational and environmental, are very similar to those used for drug analysis in biological fluids. In practice, not many laboratories would carry out both types of work.

Trace organic analysis in biological fluids is, thus, carried out for a number of important health-related purposes. Even though these differ in the information provided, the approach to developing and validating reliable, cost-effective analytical methods is very similar for each type of investigation. In all cases, despite the development of more sophisticated and sensitive analytical equipment, the major rate-limiting step is generally sample preparation. It is also the step most likely to produce the biggest source of error.

Biological samples that might be encountered include: plasma, serum, whole blood, erythrocytes, urine, breast milk, semen, saliva, cerebrospinal fluid, bile, cyst fluid, various tissues, faeces, bone, hair, teeth, expired air, etc. A particularly useful article, which deals with sample preparation procedures and how these are affected by the matrix, was published by Maickel (1984). Of the various biological matrices, the most common samples analysed are plasma, serum or urine.

Various difficulties can arise when analysing biological samples. The target analyte may often be present at low concentrations, typically ng/ml. There are often numerous endogenous compounds present, some at much higher concentrations than the analyte. In many cases, these may be similar in structure to the analyte and, therefore, the potential for interference is high. In some samples of blood, plasma or serum proteins may bind to the analyte. Analytes that are highly hydrophilic will be difficult to extract from aqueous-based samples. Analytes may be unstable to organic solvents or to changes in pH. Another potential source of error can arise if reagents, tubes, etc. give interfering peaks. There have also been examples of drugs being adsorbed onto the glassware used in analysis, giving rise to spurious results. Quite apart from the more scientific problems encountered with biological samples, sample preparation is used to extend column life, as many procedures use gas chromatography (GC) and high-performance liquid chromatography (HPLC). Direct injection of a biological matrix, such as blood plasma, can cause rapid deterioration of the column.

As a matrix, biological samples are some of the most challenging that an analyst can encounter. With a plasma or blood sample, there is often only a small volume available for analysis. There is some risk of infection, so that special procedures for the receipt and disposal of specimens is necessary. For some sample types (e.g. blood), personnel not trained in analysis are used to collect the sample. Of course, medically-trained individuals are usually needed to obtain tissue samples. In many programmes, samples will not be analysed until some time after collection, occasionally several months. It is then most important that suitable storage conditions have been established to ensure that analytes are stable until the time of analysis.

The ultimate aim of sample preparation is to provide the analyte in a form suitable for introduction into the measuring instrument. The link between the method of sample preparation and the subsequent analytical procedure cannot be overemphasised because, for example, what may be eminently suitable for a HPLC-mass spectrometry (MS)-MS procedure might be quite inadequate if applied to HPLC- with ultraviolet (UV) detection. However, even with sophisticated instrumentation, such as MS-MS, some form of sample pretreatment is usually necessary. Clearly, for most analytical methods, it is desirable to undertake the least amount of sample preparation commensurate with obtaining reliable results as cost-effectively as possible.

8.2 Techniques in common use

Many different techniques have been used to prepare biological samples for analysis. Sample preparation can be single step or multi-step, depending on the particular study being undertaken. A list of common procedures used for sample preparation or handling before instrumental analysis is presented in Table 8.1.

Table 8.1 Common sample preparation techniques

Solvent extraction	Column-switching
Solid phase extraction	Dialysis
Protein precipitation	Derivatisation
pH change	Soxhlet extraction
Liquid handling	Microwave extraction
Enzyme hydrolysis	Supercritical fluid extraction
Homogenisation	Cell disruption
Centrifugation	Filtration
Evaporation	Freeze-drying

The results of a recent survey (Majors, 1993) of trends in sample preparation are presented in Table 8.2.

Table 8.2 Procedures used in a survey of sample preparation

Procedure	%
Filtration	74
Liquid-liquid extraction	55
Centrifugation	54
Evaporation	54
Derivatisation	50
Solid phase extraction	49
Column chromatography	43
Precipitation	36
Soxhlet extraction	34
Digestion	25
Dialysis	13
Cell disruption	9

It should be noted that this survey was not restricted solely to laboratories carrying out bioanalysis but, nonetheless, it does give an indication of the most favoured procedures. The most widely-used techniques for trace organic analysis in biological fluids are described in the following sections.

8.2.1 Protein precipitation

It has long been recognised that deterioration in the performance of chromatographic columns used for plasma, serum or whole blood analysis is caused by the high concentrations of plasma proteins. If such matrices are injected directly into a gas chromatograph, they will immediately denature in the heated injection port. Denaturation can occur on the column once the proteins encounter the organic modifiers common in reversed-phase (RP) HPLC. Such denaturation can increase back pressure and modify the behaviour of the column, giving irreproducible retention times and spurious peaks, and causing a drop in efficiency.

These problems can be overcome simply by precipitating plasma proteins and removing them by a quick, low-speed centrifugation step. Protein precipitation can be carried out by a number of methods, such as: use of water miscible solvents (acetonitrile, methanol, ethanol, acetone); use of acids, such as trichloracetic acid (TCA), perchloracetic acid (PCA); use of bases, such as ammonium sulphate; heating; and use of chaotropic reagent (urea). Of these, the most convenient and widely-used approach is the use of water miscible organic solvents. These can be evaporated to dryness provided the analyte is nonvolatile, allowing reconstitution in, for example, the HPLC mobile phase. It has been found necessary to have a five-fold excess of organic solvent to ensure efficient precipitation of protein. TCA and PCA remove proteins efficiently but are not so compatible with the requirements of the next step and cannot be easily removed. Few approaches use heating or chaotropic reagents. With each of the above-mentioned procedures, there is the possibility that the analyte may become physically trapped in the protein as it precipitates.

Protein precipitation is not as rigorous a clean-up as liquid-liquid extraction (LLE) or solid phase extraction (SPE), as many potentially interfering compounds will be soluble in the organic solvent or acid as the analyte must be. Until the advent of HPLC-MS-MS, sample preparation that involved only the simple procedure of protein precipitation was of limited applicability, because the analytes that were to be determined were not usually present in sufficient concentration. One of the problems of the simple protein precipitation approach was that, in general, it resulted in the dilution of the sample. In addition, when organic solvents were used, the high eluotropic strength of the sample meant that it was not possible to compensate by injecting larger amounts onto the column. The need to remove the organic solvent and reconcentrate the sample meant that the attractiveness resulting from the simplicity of the approach was much diminished. However, for those limited circumstances where analyte concentrations are relatively high, protein precipitation may represent the method of choice. In addition, where a specific and sensitive detection system is available, e.g. HPLC-MS-MS, such a simple approach will often provide clean enough extracts for direct analysis. Indeed, where HPLC-MS-MS is the method of choice for bioanalysis, protein precipitation and direct injection, so called 'dilute and shoot', procedures are usually the first to be investigated in a method development programme.

However, as indicated above, a mass spectrometric assay is not always needed and protein precipitation can prove adequate when higher concentrations are expected in plasma or serum (μg/ml). For example, with piperacillin and ceftazidine, an HPLC-diode array detector (DAD) was used after protein precipitation with methanol (Campanero et al., 1997b). It was also found that, after precipitation with acetonitrile, tinidazole and metronidazole could be measured by HPLC-UV without difficulty (Stevenson, unpublished).

8.2.2 Liquid-liquid extraction

Where simple methods, such as protein precipitation, are not practicable, some means of extraction is usually employed. Such extractions can be used for analyte concentration, removal of interferents or both. Despite the development of alternative methods of sample preparation, such as SPE, liquid-liquid (solvent) extraction remains a very popular technique in the bioanalytical field because it is simple, easily understood, reliable and readily implemented. Liquid-liquid extraction (LLE) methods rely on the fact that many drugs possess a sufficiently lipophilic character to allow partition between an aqueous and an immiscible organic phase. The mixing is typically carried out using a vortex mixer, a roller mixer, or apparatus that constantly inverts tubes. Manual procedures are carried out in test tubes, often with stoppers, not in separating funnels. A quick, low-speed centrifugation step is used to separate layers, and the organic solvent is removed manually into a second tube.

A wide variety of solvents is available, ranging from nonpolar solvents, such as hexane or heptane, to more polar solvents, such as dichloromethane (DCM) or ethyl acetate. Solvents can be arranged in order of polarity, so that a rational choice can be made. Solvent selection is important. The main criteria for selection are that the solvent: is immiscible with water; has optimum polarity to match that of the analyte; is volatile if it is to be evaporated to dryness; or is compatible with the next stage of analysis if it is to be injected directly into a chromatograph; and is preferably of low toxicity and environmentally-friendly. In bioanalysis, one approach is to use the least polar solvent that will still give quantitative recovery. Of course, when developing a method, it is not only the recovery of analyte that is important but consideration of the amount of interfering material also being extracted from the biological matrix. With this approach, it is quite common to use a volatile solvent, such as diethyl ether. This has the advantage that it can easily be evaporated to dryness allowing preconcentration of analyte (provided the analyte is relatively nonvolatile). It is usual to choose a solvent lighter than water so that the upper layer can be removed but, if a solvent heavier than water is desired, chloroform or DCM have been used. Traditionally, a wide range of solvents has been employed but this is less desirable now as some solvents will be unsuitable on the grounds of their toxicity or harmful impact on the environment.

As many analytes are ionisable, they can exist in both the ionised and nonionised form. Which form predominates depends on the pH of the aqueous phase and the pk_a of the analyte. This is shown for a model acid and base below:

$$RCOOH \longleftrightarrow RCOO^- + H^+$$
$$\text{nonionised} \qquad \text{ionised}$$

$$BH^+ \longleftrightarrow B + H^+$$
$$\text{ionised} \qquad \text{nonionised}$$

$$\text{Acidic} \longleftarrow pH \longrightarrow \text{basic}$$

As the nonionised form is more likely to partition into the organic phase, a high pH will favour extraction of a base and a low pH will favour extraction of acids. One possible problem with LLE from plasma or serum is the formation of a single layer emulsion when mixing vigorously. The most common methods to overcome this problem are: adding salt to the aqueous phase; heating or cooling the tube; filtering through a glass wool plug or filter paper; and adding a small amount of a different organic solvent (Majors, 1993).

With LLE of biological samples, most common procedures would take a small volume of sample, typically 0.5 or 1 ml, and extract with 5 or 10 ml of organic solvent. This phase volume ratio facilitates quantitative extraction into the organic solvent even for compounds whose partition ratio is not particularly favourable.

More extensive sample clean-up can be obtained by using a procedure with back-extraction. This can be achieved for analytes containing an ionisable group by choosing a pH to suppress ionisation and extracting into an organic solvent. Once this has been done, further buffer is added at a pH to promote ionisation and, thus, to extract the analyte back into a clean aqueous phase. After phase separation, if compatible with the instrumentation, the sample is then injected into the measurement apparatus. Alternatively (and more commonly), the pH of the aqueous extract is altered by the addition of a buffer, acid or base, to suppress ionisation of the analyte, and extraction into an organic solvent is once more performed. The solvent is evaporated to dryness and redissolved in a small volume of suitable solvent. Such a multistep extraction is time-consuming but may produce clean extracts for HPLC, gas-liquid chromatography (GLC), etc. Losses may occur at each stage and transfer of phase volumes may not be exact, so the use of suitable internal standards is recommended to compensate.

One less commonly used approach to LLE is ion-pair extraction. With this approach, a pH is chosen to promote ionisation of the analyte and a reagent (the pairing ion) of the opposite charge is added to form a neutral complex. The pairing ion contains hydrophobic groups to enhance the extraction of the ion-pair complex into the organic solvent. The equation for ion-pairing is presented below:

$$B^+_{aq} + P^-_{aq} \longrightarrow [B^+\ P^-]_{org} \tag{8.1}$$

where, B is the basic drug to be extracted, and P is the pairing ion.

The equilibrium constant (K) for the reaction is given by:

$$K = \frac{[B^+P^-]_{org}}{[B^+]_{aq}[P^-]_{aq}} \tag{8.2}$$

The distribution of B between the phases is represented by:

$$\frac{[B^+P^-]_{org}}{[B^+]_{aq}} \text{ which } = K[P^-]_{aq} \tag{8.3}$$

The extraction of B can, thus, be regulated by the concentration as well as the type of pairing ion. For basic compounds, typical pairing ions include heptane and octane sulphonic acids and sodium lauryl sulphate. For acidic analytes, quaternary ammonium salts are used, such as tetrabutyl ammonium chloride. The ion-pair approach is useful for highly polar analytes that are difficult to extract at any pH via the ion-suppression approach described previously.

It is often, and quite incorrectly, stated that LLE methods are not amenable to automation. Such statements are usually made as a justification for adopting an automated SPE method. In fact, in our experience, the automation (see Section 8.2.12) of LLE using Zymark robots is remarkably simple, and results in a very robust and reliable extraction method. Indeed, despite the development of alternative methods of sample preparation, many new procedures involving simply LLE followed by evaporation to dryness and redissolution in HPLC mobile phase are still being reported. Recent examples include methods for verapamil in blood, liver and kidney by HPLC-fluorescence (Negrusz *et al.*, 1997), and lomefloxacin, fenbofen and felbina in plasma by HPLC-UV (Carlucci *et al.*, 1996). An interesting example, in which LLE and (on-line) SPE are compared for the analysis of psilocin, is provided by Lindenblatt and co-workers (1998), who concluded that both methods gave essentially the same results (although on-line extraction was more convenient).

8.2.3 Solid phase extraction

SPE is becoming the method of choice in many laboratories carrying out bioanalysis, particularly within the pharmaceutical industry. SPE involves passing a liquid sample through a bed of sorbent, usually held in a small plastic syringe cartridge. The aim is to selectively retain the analyte, and then to elute it in a small volume of clean extract. A wash step can be incorporated if necessary, as is common in biofluid analysis. An idealised SPE sequence for biological fluid might involve: (i) activation of the sorbent; (ii) removal of excess activation solvent; (iii) conditioning of the sorbent; (iv) application of the biological sample; (v) washing the cartridge with a solvent not quite strong enough to elute the analyte, to remove interferents less strongly bound than the analyte; and (vi) elution of the analyte in a small volume of a solvent, just strong enough to elute the analyte but leaving more strongly bound interferents on the column.

The extract may be suitable for direct injection into a GLC or HPLC, etc., or is more likely to be evaporated to dryness and redissolved in a smaller volume of solvent compatible with the requirements of the instrumental analysis. The cartridges are most commonly used with a multi-place vacuum manifold to optimise the flow through the sorbent. Positive displacement by gas pressure at the top of the bed is also possible.

One of the major perceived advantages of SPE is that it is easily automated, and a number of manufacturers have developed systems that enable either off-line or on-line extraction (see Section 8.2.12).

A wide and growing range of phases has been used for bioanalysis, particularly drug analysis. Examples of the sorbents used are presented in Table 8.3.

Despite this wide range, the most common phases for aqueous biological samples are C2, C8 and C18 bonded silica. SPE is a form of low efficiency chromatography and the mechanisms are, in principle, the

Tabel 8.3 Examples of sorbents used for solid-phase extraction (SPE) of biological fluids

Nonpolar	Polar
C18	Cyanopropyl
C8	Diol
C2	Silica
Cyclohexyl	Aminopropyl
Phenyl	N-propylethylenediamine
Carbon	

Cation exchange	Anion exchange
Benzenesulphonylpropyl	Diethylamino
Sulphonylpropyl	Trimethylaminopropyl
Carboxymethyl	

Other
Mixed mode
Porous polymers
Phenyl boronic acid
Size exclusion
Affinity

same as for HPLC. With SPE cartridges, the particle size is $\sim 40\,\mu m$ to keep back pressure low, and pore size is $\sim 60\,A$. The capacity for mass of analyte is greater than HPLC, typically 5 mg for a 100 mg cartridge. Sorbent bed mass can vary from 50 mg to 1 gm. In addition to cartridges, a variety of other SPE formats are available, including disks (polytetra fluoroethane (PTFE) or glass fibre) in which the SPE phase is embedded. These disk-based systems are used in exactly the same way as cartridges, but may be particularly well-suited to small volume samples and require much smaller volumes of eluent.

Once the phase type has been selected, optimisation of the extraction conditions for a particular analyte usually focuses on the choice of washing and elution solvent. For aqueous biological samples, this usually means a reversed phase type as the first approach; variation of solvent polarity and pH receive initial consideration. Separation is supposedly based on hydrophobic interactions. However, residual silanols can act as polar sites or as anion-exchangers [SiO^-]. A more refined understanding of the mechanisms of separation has shown that the residual silanols may play an important role in many extractions (for example, see Ruane and Wilson, 1987; Roberts *et al.*, 1989; Martin *et al.*, 1996), even though this may not have been understood when the method was proposed. These 'secondary interactions' may, in fact, be the primary interaction, particularly with basic drugs . As the number of residual silanols may vary from batch to batch, an extraction based unknowingly on this mechanism would be prone to batch-to-batch variation.

As an ion-exchange mechanism may operate for basic drugs, it follows that simple water-organic solvent mixtures may be relatively ineffective eluents for this type of analyte, as they fail to break the relatively strong ionic interaction. To overcome this, conditioning of cartridges with solvents containing, for example, Na^+ or K^+ ions may improve extraction of basic drugs by blocking residual silanols on reversed-phase materials. Alternatively, modifiers, such as triethylamine, can be used in the eluent to compete with the analyte for the silanophilic sites. Trifluoroacetic acid, or some other acidic modifier, can also provide an effective means of overcoming this type of interaction and ensuring efficient recovery. Column conditioning may, thus, exert an important influence on analyte extraction. Recently, a base deactivated C8 material has been developed, which is less prone to interactions of this type should the analyst wish to avoid them (Martin *et al.*, 1995). Other materials that also achieve this end by restricting access to silanols have been developed, including so-called 'shielded' phases. These may have advantages under some circumstances (although they are not without problems) (Martin *et al.*, 1996). It is worth noting, however, that this type of 'mixed mode' extraction mechanism involving silanol interactions can prove very useful to method development in enabling selectivity to be built into the method. As a result, special mixed mode phases, using either blended mixtures of phases or silica with multiple functionality, have been developed specifically to exploit these advantages in a controlled way.

In theory, many factors need to be optimised (see Table 8.4); however, in practice, it is often the case that time does not allow such a rigorous approach and, indeed, it is often unnecessary. Many SPE procedures are based on: activation/conditioning with methanol and buffer; sample loading; washing with a suitable mixture of methanol in buffer; elution in methanol.

Reports of SPE followed by HPLC are becoming ever more common in the literature, with recent examples including: amphetericin B in plasma and sputum using C2 extraction and then HPLC-UV (Campanero,

Table 8.4 Factors influencing optimisation of solid-phase extraction (SPE)

Choice of phase type
Phase weight
Particle size
pH of sample
Dilution of sample
Cartridge conditioning solvent
Wash solvent polarity, pH
Elution solvent polarity, pH
Flow rate

1997a); a C2 end-capped phase was selected as giving the best result for the extraction of racemic citopram and metabolites from plasma by Carlsson and Norlander (1998); leukotriene metabolites in cell culture samples using on-line extraction on C18 and then HPLC-DAD (Heintz et al., 1997); tetracycline antibiotics in serum using Oasis HLB extraction and then HPLC-UV (Cheng et al., 1997); fluvastatin enantiomers in plasma by automated SPE on C2 columns and then HPLC on a chiral stationary phase (Toreson and Eriksson, 1997); and melatonin in plasma after C18 extraction and HPLC-fluorescence (Kulczykowska and Iuvone, 1998). Coupling the very high resolution achievable with capillary electrophoresis with the specificity and sensitivity of MS for the analysis of a candidate drug in plasma still required SPE with C2 columns (Paterson et al., 1997).

It is worth noting that, where the analysis is particularly difficult, it may be necessary to link a variety of sample preparation techniques together in order to achieve the desired result. An example of this is provided by the analysis of the anabolic steroid, methyl testosterone, in hair. This required a multistage clean-up involving LLE, SPE, HPLC and collection of fractions from HPLC for detection by enzyme immunoassay (Gleixner et al., 1997).

Because of the popularity of SPE, several methods have now been published which employ the technique and many manufacturers provide extensive bibliographies describing these methods. This information is a useful starting place in method development for new compounds, and may provide a ready source of predeveloped extraction schemes. A further useful source of information on SPE is a recently published book on the subject by Thurman and Mills (1998), which also provides details of some methods for drugs and pesticides, etc.

In an alternative approach, the SPE phases can also be used in a technique called 'matrix solid-phase dispersion', where the phase is homogenised with the sample rather than used in a column or as a disk. This method has been used for tissue samples, such as liver and muscle. It produces a homogeneous and rapid blending of the biological matrix compared with the column or disk method.

One area of SPE currently attracting a lot of interest is the use of antibodies or molecular imprinted polymers as highly selective SPE columns. Biological antibodies can be covalently bonded onto silica or controlled pore size glass, while retaining the specificity of the antibodies. Protocols have been developed to selectively retain and then elute analytes in small volume fractions, in the best cases as little as 1 ml. Immunoextraction procedures have been developed for the phenylurea herbicides, chlortoluron and isoproturon, and the drugs, morphine and clenbuterol. All of these, basically, utilise a very similar protocol, i.e.

loading the sample onto the column at neutral pH in phosphate-buffered saline (PBS), and then eluting at low pH (typically pH 2) together with 40–50% ethanol or methanol (Rashid et al., 1996; Martin-Esteban et al., 1997; Stevenson et al., 1998; Shahtaheri et al., 1998).

In all four examples, it was demonstrated that the retention was due to the antibody to the compound of interest. Retention on a column containing antibodies to other compounds did not give the same retention and elution in a single fraction. In the case of the two drugs, a suitable protocol was developed for morphine in urine or clenbuterol in plasma. The method for clenbuterol in plasma gave clean traces, but in a preliminary validation of the method using spiked samples, better day-to-day reproducibility was obtained if the plasma proteins were precipitated first. As antibody columns could be considered expensive, experiments to examine their reusability were undertaken. These suggested that they could potentially be reused 20 times or more without loss of antibody-binding capacity. As analytical instrumentation can detect lower and lower levels, such considerations may prove unimportant. The main reason for using large amounts (typically up to 200 µl of antibody) was to increase the capacity of the column to hold analyte. Much lower volumes of antibody can be used if it is necessary to retain only small amounts of analyte. As a very broad approximation (because they vary widely) 10 µl of antibody could retain 10 ng of analyte. If a clean chromatographic trace were obtained such that no further sample preparation was needed, it would often be unnecessary to reuse the columns on the grounds of cost. In that case, 1 l of antisera would provide for 100,000 assays.

An interesting variation on the development of selective solid phases is the use of molecular imprinted polymers (MIPs) to mimic the behaviour of biological antibodies. These again selectively bind to a particular analyte present in a complex mixture of other compounds. This would allow the production of phases in the synthetic chemistry laboratory rather than relying on the less reliable biological systems. The simplest approach to preparing a MIP uses a monomer, such as methacrylic acid, and a cross-linking agent, such as ethylene glycol dimethacrylate, to polymerise around a template, which is the analyte molecule. These are dissolved in a solvent, such as acetonitrile, and an initiator, such as 2,2′azobis-(2-methylpropionitrile), is added. The reaction is allowed to proceed for 16 h at a temperature of about 60°C . At the end of this time, a solid mass of polymer has formed. The polymer is removed and ground into particles with a mortar and pestle. It is then extensively washed to remove the analyte template. The resulting polymer hopefully has cavities containing a rigid imprint of the analyte. Optimisation of loading, washing and elution of the solvent allows selective extraction of the analyte molecule (Mayes and Mosbach, 1997; Sellegren, 1997).

Early SPE experiments with MIPs have suggested that it is rather difficult to extract all of the template molecule. This leaches out with time and would, of course, interfere with subsequent use of the material as an analytical SPE phase. There is some evidence to suggest that the analyte leaches out with time, as if some analyte is slowly migrating to the surface and cannot be washed out easily during the preparation of the phase. Early reports (Rashid et al., 1997) also suggest that the MIP based supports are not as specific as phases based on biological antibodies. This could be used to some advantage, if phases are prepared to a structural analogue of the analyte of interest. Potentially, classes of compound could be extracted or if analyte leaching is a problem an analogue is used to make the imprint and this would be separated by chromatography. An alternative type of MIP has been prepared by synthesising a derivative of the analyte that contained a polymerisable functionality, for example vinylphenyl boronate. After polymerisation, the analyte was hydrolysed and the remaining polymer used as a chromatographic phase (Wulff, 1995).

If specific polymers can be synthesised without problems of analyte template removal, these would undoubtedly prove useful for sample clean-up. Currently, the literature on the use of MIPs for sample preparation contains few examples of their use in SPE, but these can be expected to increase rapidly as more groups investigate applications of this type of technique.

8.2.4 *Homogenisation/hydrolysis of solids*

The determination of trace organics in tissue usually requires mechanical destruction of the tissue and release of the analyte into a homogenate. Mechanical destruction procedures include high speed, rotating, sharp blades and ultrasonic probes. The homogenate could be an aqueous buffer or an organic solvent depending on the lipophilicity of the analyte. Homogenisation is usually followed by hydrolysis of tissue proteins by heat, strong acid, alkali or the use of enzymes. Alternatively, protein precipitation is carried out as described previously. Analytes may not be stable to strong acid or base, and enzyme hydrolysis offers an alternative. Enzymes commonly used include a preparation obtained from *Subtilisin carlsberg* and other proteases.

8.2.5 *Hydrolysis of conjugates*

Many drugs (and other trace organics) undergo phase I and phase II metabolism. Phase I metabolism involves the addition or removal of a proton or a functional group, such as hydroxyl or methyl, usually producing one or more metabolites that are more polar than the parent compound. Phase II metabolism involves conjugation of the compound and/or metabolites with glucuronic acid, sulphuric acid or glycine. These conjugates are much more polar than the original analyte and can be difficult to extract or to analyse directly. Many analytical procedures involve conjugate hydrolysis back to the original drug or metabolite, etc., before subsequent extraction and analysis of the aglycone. This is usually carried out by heating with acids or incubating with glucuronidase, sulphatase or a mixture of the enzymes. Problems can arise with ester glucuronides, as these have a propensity to rearrange under mildly alkaline conditions to give products that are resistant to enzymic hydrolysis. In such circumstances, providing the aglycone is stable, it is better to use alkaline hydrolysis to cleave the ester bond rather than enzymes.

8.2.6 Dialysis

Dialysis involves the separation of higher molecular weight compounds from those of lower molecular weight using their differential permeation rates through membranes. Plasma contains high concentrations of high molecular weight compounds, such as proteins, and they can be separated from drugs, etc. to reduce problems with HPLC columns. The analyte transfers from the donor solution into a recipient solution. The process is diffusion controlled and is, therefore, quite slow. The analyte is transferred to a dilute solution and requires further preconcentration. The greater the concentration difference between donor and recipient the faster the diffusion of the analyte. With automated equipment, such as automated sequential trace enrichment of dialysates (ASTED), the recipient solution is constantly replaced in order to retain a concentration gradient. The diluted recipient solution is passed through a column, where analyte is retained and preconcentrated. The method can be completely automated (Cooper *et al.*, 1994).

8.2.7 Column-switching

Column-switching has been employed to good effect in bioanalysis. The technique involves the coupling together of two or more columns (usually HPLC) and the use of a switching valve to divert the eluent from one column to the other, using a different solvent. In its simplest form, the general aim is to send only the fraction of interest from the first to the second column. Column-switching techniques have been used to assay biological samples without further clean-up and the whole process can be automated. Accurately timed switching and highly reproducible retention times are required. As more than one solvent is used (one which will not elute the analyte from the first column and then one that will) these must be compatible with each other.

There are several alternatives in column-switching techniques, such as back-flushing of columns and the use of more than two columns. Stationary phases can be of the same or different types. One interesting recent application of column-switching has been for chiral separation, where flow from an achiral column is switched to a chiral column.

Recently, column-switching has been reported for the determination of parathion and metabolites in serum using RP-8 as the first column and switching to octadecyl silica (ODS). The serum was diluted with phosphoric acid and injected directly. Detection was by HPLC-UV (Lee *et al.*, 1997). A new approach has proposed the use of a restricted access column as the first column, allowing direct injection of biological fluids and then switching to conventional columns (Yu and Westerlund, 1997).

8.2.8 Restricted Access Media (RAM) columns

One method of avoiding or minimising sample preparation with biological fluids is the use of RAM columns. These can be used in the single column or in column-switching modes. These columns allow the analysis of low molecular weight compounds after direct injection of the biological matrix. High molecular weight molecules, such as proteins, are excluded from access to the adsorptive sites of porous supports and elute in the void volume. Small solutes, such as drugs, can reach the adsorptive sites and are retained to some degree. The size exclusion of macromolecules is achieved using either a chemical or a physical diffusion barrier, depending on the type of column used. If direct injection of plasma is made, the mobile phase must not precipitate the plasma proteins. A number of commercially available materials and applications were reviewed (Boos and Rudolph, 1998a and b). Both reversed phase and mixed mode retention is possible.

8.2.9 Volume reduction

Several of the above-mentioned procedures involve extraction into an organic solvent. Many methods require the reduction or complete removal of solvent before introducing samples into, for example, HPLC or GC. In some cases, this is simply to allow reconstitution in HPLC mobile phase to allow complete compatibility with the system, thereby avoiding a large injection peak. In other cases, it may simply be done to gain sensitivity by redissolving in the lowest possible volume. With organic solvents, the most common procedure is to blow oxygen-free nitrogen across the surface of the solvent, with or without gentle heating. If larger volumes of solvent are to be removed, special apparatus, such as a Kuderna-Danish concentrator, are used. Large volumes of solvent or water can be removed on a rotary evaporator, with a vessel connected to allow collection of the distillate. For aqueous samples, freeze-drying can also be used. When evaporating to dryness, care must be taken not to lose analyte by evaporation or by irreversible adsorption onto glassware.

8.2.10 Derivatisation

In many analytical methods, it is necessary to derivatise the analyte before it is measured or extracted. This is particularly common in GC but is also carried out for HPLC and other types of bioanalysis. With GC, the major reason for derivatisation is to increase the range of compounds that are volatile enough to be analysed by the technique. This was particularly the case before the advent of HPLC and capillary electrophoresis (CE) but also remains the case today.

The usual approach is to look for polar functional groups and to 'cap' these by reacting with a suitable derivatising agent. Problems caused by polar functional groups include: tailing peaks; adsorption onto the stationary phase, if this is not uniformly coated; difficulty with extraction. Apart from the reactions performed to improve volatility, derivatisation is also carried out to improve detection of analytes. A functional group that responds to a particular detector is added so that even if the analyte does not respond, for example to an electron capture detector (ECD), the derivative does. Again, this is much more commonly used with GC than with HPLC, but HPLC methods using this approach are not unusual.

A further reason for derivatisation is the need to prepare diastereoisomers for separation of enantiomers. These can often be separated on achiral columns. This approach is still extremely common both in GC and HPLC despite the advances in commercially available chiral columns. For example, in a recent application, the analysis of ibuprofen enantiomers in plasma and urine required LLE, SPE and derivatisation; the later step to give diastereoisomers, so that separation was achieved on an achiral column by HPLC with fluorescence detection (Tan *et al.*, 1997).

For bioanalysis, derivatisation can be carried out before or more commonly after extraction of analyte from the biological matrix. Derivatisation after extraction is more common, as many reactions require nonaqueous conditions. In the case of HPLC, reactions could be pre- or post-column. Functional groups commonly derivatised include:

carboxylic acid, —COOH; sulphonic acid, —SO$_3$H; hydroxyl —OH; amines (primary, —NH$_2$; secondary, —NHR; and tertiary, —NR$_2$); ketones, =CO; aldehydes, —CHO; thiols, —SH; and nitrosamines, N—NO.

The exact reaction used will depend on the functional group to be derivatised and on the purpose of the derivatisation. Derivatisation reactions should ideally be quantitative, rapid, and allow excess reagent to be removed easily. An excellent source of procedures is the book by Blau and Halket (1994) (see Bibliography).

Reactions in common use are: esterification, e.g. with methanol/HCl, alkyl halides, diazomethane, boron trifluoride/methanol; acylation, e.g. with acetic anhydride/pyridine, trifluoracetyl chloride; silylation, e.g. with N,O-(bistrimethylsilyl) acetamide (BSA), bis(trimethylsilyl) trifluoroacetamide (BSTFA), trimethylsilyl (TMS); alkylation, e.g. with alkyl halides; Schiff base formation, e.g. with phenyl hydrazine; and cyclic compound formation, e.g. with aldehydes and ketones. Most of these reactions have been used for GC, and recent examples from the literature include the use of derivatisation for this purpose. Thus, the use of SPE followed by derivatisation with BSTFA, then GC-MS is described (Jönsson and Åkesson, 1997) for the determination of 5-hydroxy-N-methyl-2-pyrrolidone and 2-hydroxy-N-methylsuccinimide, both metabolites of the widely-used solvent N-methyl pyrorolidone. GC (but with ECD) after LLE then derivatisation with BSA was also used for the determination of the drug, temazepam, in plasma (Escoriaza *et al.*, 1997). An interesting example of combining SPE with *in situ* derivatisation is provided by Yu and co-workers (1998). Thyreostatic drugs were extracted from a muscle matrix onto an anion ion-exchange resin and were then methylated with methyl iodide in acetonitrile prior to GLC.

Derivatisation in HPLC was normally used to produce an ultraviolet-absorbing or fluorescent derivative to enable detection of an analyte that did not possess good detection capability. Perhaps the best example of this approach is the determination of naturally occurring amino acids, most of which do not posses chromophores. Reagents include benzoyl chloride and phenyl isothiocyanate. Some compounds (tamoxifen for example) can be converted to fluorescent derivatives by irradiation with UV light.

Practical considerations, such as the need for special reaction vessels, the toxicity of reagents, the need for nonaqueous conditions, the possibility of reaction with endogenous compounds, the need to carefully control reaction conditions, etc., mean that derivatisation is usually avoided if possible. Nonetheless, it is still common for analytes where selective detection is required and polar groups cause problems.

8.2.11 Headspace analysis

The determination of volatile organics in biological samples can be achieved by headspace analysis. In this process, the sample is placed in a sealed container with a rubber septum in the cap. Samples and standards are allowed to equilibrate at a set temperature. Instead of taking a liquid sample, a large volume (typically as much as 1 ml) of the vapour above the sample is taken. The concentration of analyte in the vapour above the liquid will be in direct proportion to the concentration in the fluid. This approach is normally used with GC, as analytes need to be relatively volatile. It has the considerable advantage that almost all of the likely interferents in a biological sample are not volatile enough to be present at high concentration in the headspace. Volatile components, such as propellants for drug delivery systems, solvents, etc., will reach concentrations high enough to be measurable. Headspace analysis is in common use for the determination of volatile organics in water, and has been proposed for biological monitoring of exposure to certain volatile compounds in the workplace.

A new technique in this area that shows much promise is solid-phase microextraction (SPME), whereby a fibre, coated with a suitable sorbent, is used to sample the volatiles. The whole fibre is then introduced into the chromatograph and the analytes are desorbed in the usual way. The current practice of SPME has recently been comprehensively reviewed in a book by Pawliszyn (1997). This technology is also beginning to be applied to analytes in aqueous samples and biofluids.

GLC is still the method of choice for certain classes of analyte, such as local anaesthetics and other volatiles, and for these classes methods like headspace analysis, and its combination with SPME, are very useful. A number of recent examples of applications for biofluids have been published. For example, the application of headspace analysis and GC-flame ionisation detection (FID) has been used to measure formaldehyde in urine (Tashkov, 1996). Local anaesthetic gases in blood have also been measured by GC-FID following direct immersion of a SPME probe after precipitation of plasma proteins with perchloric acid (Kumazawa *et al.*, 1996). The analysis of local anaesthetics in blood using headspace SPME and GC-MS has recently been described (Watanabe *et al.*, 1998). SPME has also been used to prepare samples of 21 amphetamine-related compounds in urine for simultaneous GC-MS determination (Battu *et al.*, 1998), as well as amphetamine and methamphetamine in human hair by headspace SPME and GC-nitrogen-phosphorus detector (NPD) (Koide *et al.*, 1998).

8.2.12 Automation of sample preparation

Automation of methods provides a means of improving sample throughput and removing the tedium of routine bioanalysis. What exactly constitutes automation is sometimes difficult to define and, for example, the column-switching technique described above can be considered as such. In this section, the automation of extraction methods using robotic techniques and similar dedicated instrumentation will be considered.

Many bioanalytical methods that began life as manual techniques are eventually converted into automated ones. In some cases, where it is clear at the outset that a new method will be required to analyse a large number of samples, an automated method may be developed from the start. The process of automation over the last 10 yrs can be seen as one of expansion or continuation of the early advances made by pieces of equipment, such as multi-tube mixing or shaking devices, multi-tube liquid handling systems for dilution of large numbers of samples simultaneously or autosamplers for HPLC or GC, and systems such as column-switching. Such systems can be considered as automating a part of the analytical process. Robotic automation has taken these earlier steps further by combining such processes or equipment together into more multifunctional systems, for example an autosampler may now be a part of a fully-automated sample preparation and analysis system rather than the only piece of automation in a primarily manual method. Robotic automation has attempted to automate whole methods or large parts of them by combining the automation of the individual processes.

The earliest forms of automation tended to be autosampling machines. As with the development of most forms of automation, target areas are any rate-limiting steps in a process and areas containing a high level of repetition. Hence, autosamplers were designed to eliminate the need for an analyst to sit by an instrument manually sampling at theoretically routine intervals over long periods of time. The development of the autosampler, whilst it did not eliminate a rate-limiting step, certainly freed up the analyst's time, and also led to better analytical results through greater precision of sample introduction. However, this time was then diverted towards sample preparation in order to provide samples for the autosamplers.

The process of robotic automation of sample preparation was a much bigger step than that of automatic sampling of vials, since not only did a number of different processes require automation, but these processes then had to be strung together in a sequence to give the same or an improved end-product to the analyst.

Both LLE and SPE are amenable to robotic automation and, contrary to the general perception, our experience of automating LLE has been very positive. Indeed, the automation of this process using Zymark robots, initiated in our laboratories over 14 yrs ago, has been so effective that it would be very unusual to find an analyst carrying out such work manually. In some ways, we have found automated LLE methods rather easier than those developed using SPE. One problem with SPE is that samples containing plugs of protein, or other particulates can cause the frits to clog preventing sample loading. A human analyst can see this occurring and take the necessary corrective action, but an automated system cannot (this type of problem does not affect LLE methods to anything like the same extent). The transfer of a manual method to a robotic one may, therefore, not be as straightforward as one might expect, as it may be that extra filtration steps, etc., may be needed to compensate for the absence of a human operator.

The benefits of automating bioanalytical sample preparation are, however, very great. Thus, current systems like the Zymark robots are capable of running over 24 h at a rate of one sample every 5 min with very little maintenance. The rate-limiting step of sample extraction by the development of this technology has, therefore, been eliminated. Considerable benefit has been seen as a result of automating the extraction process. Analyst time is freed to provide better use of an increasingly limited resource. In addition, increased reproducibility of runs due to higher levels of accuracy and precision usually result from a well-developed, automated method. By eliminating analyst errors, a benefit which noticeably increases in the case of complex assays or 2–3 day assays, there are less failed runs, etc. Unstable compounds that may deteriorate with standing are handled much more effectively, since the preparation of each sample always takes the same length of time, and the sample can then be injected directly into the chromatograph. A manual process, by comparison, may involve processing aliquots of a batch of samples over a period of time followed by mixing and centrifuging as a single batch; hence, the standing time for the first aliquot would be greater than the last. It is also worth noting that automated systems would be of great benefit when working with light sensitive compounds (as the systems can be operated in total darkness) or hazardous materials/samples (e.g. human immunodeficiency virus positive (HIV+) plasma, etc.).

Processes that have been combined in robotic sample preparation include: liquid handling; vortex and inversion mixing; filtration; protein precipitation; SPE; evaporation; centrifugation; redissolution; and fraction collection. We have also incorporated autosampling and the HPLC analysis into the procedure, so that the whole preparatory and analytical process from raw sample to result is automated.

Our success in the robotic automation of LLE has been such that we have tended to focus primarily on the use of this technology for such methods. Whilst we have used robots for SPE, a better option for this type of sample preparation is probably to employ one of the many dedicated SPE workstations that are now available. For large sample numbers and high throughput assays, the use of 96-well plate technology is very effective. These are examples of the use of this type of system in the literature, with a useful recent example that combines it with a 'generic' HPLC gradient and MS-MS detection (Ayrton *et al.*, 1998). SPE phases are now available from most manufacturers in this format and the transfer of a 'traditional', cartridge-based SPE method to the 96-well format should not cause major problems. Such systems can be particularly useful when fast turnaround times are required, e.g. in early 'rising dose' clinical studies, and to cope with the high sample throughput requirements of large scale clinical trials. For shorter runs or where more flexibility is required, the use of instrumentation that can cope with conventional SPE cartridges may be more appropriate, and a variety of such devices are available from manufacturers.

Whichever route to automating a bioanalysis is chosen, it is worth remembering that the investment of time required to ensure that the resulting method is robust is probably only worth undertaking if a number of criteria are met. Thus, such methods would generally best be used for either 'generic' extractions (where the same methodology could be applied to a wide range of analytes and, thus, a large number of samples), or to compounds which will be analysed in large numbers.

8.3 Applications

A number of applications have been used as illustrations in the preceding text and, given the diversity of such applications, a comprehensive review is not practicable in a chapter of this size. However, the scientific literature contains many references to the analysis of drugs, etc., in biological fluids, especially plasma and urine, and clearly these will contain an enormous amount of information on sample preparation. A very useful source of methods is available in the book series Methodological Surveys in Biochemistry and Analysis, many of which are devoted to the bioanalysis of drugs (see Bibliography). This series has descriptions of bioanalytical methods covering several hundred drugs, and is far more than a list of recipes. Similarly, as indicated in the section on solid-phase extraction, a wealth of information on applications in bioanalysis is available from the manufacturers.

8.4 Factors influencing choice of method

The exact approach taken to prepare biological samples for analysis will depend on many factors, so it is difficult to prescribe a definite approach. With sample preparation, one fact is usually true, namely that the analyst does not really want to carry out this step. Sample preparation is only undertaken because the consequences of not doing so would cause problems. The real desire is to carry out the minimum amount of sample preparation commensurate with achieving the sensitivity, specificity and sample throughput needed for a particular study. Some of the factors influencing the amount of sample preparation and choice of approach are: the chemical nature of the analyte(s); the complexity of the biological matrix; the physical state of the matrix; the amount of sample available; the likelihood of compounds of similar chemical structure to the analyte being present in the matrix; the chemical and biological stability of the analyte; the specificity of the end-step; the sensitivity of the end-step; the detection levels required; and the accuracy and precision required.

Sample preparation, thus, tends to be application driven. Other less scientific considerations will also be involved, such as the availability of particular techniques and expertise, individual choice and how quickly results are needed. With biological samples, highly specific detection systems, such as the electron capture detector for GC or the fluorescence detector for HPLC, and MS for both, can allow minimal sample preparation. If levels to be determined are high (e.g. µg per ml rather than pg per ml) then less sample preparation is needed.

On the other hand, when the concentrations to be determined are very low and no specific, sensitive detector is available then a multistage procedure might be necessary.

8.5 Future prospects

Increasingly, the trend is towards the use of HPLC-MS and MS-MS as the end-points in bioanalysis. These techniques can provide exquisite sensitivity and specificity and high sample throughput. The particular advantage of these methods is that sample preparation can often be minimised. It is, however, difficult to underestimate the importance of sample preparation, even in an age where the availability of MS as an HPLC detector is causing a revolution in bioanalysis. There are still many instances where concentration of the analyte must still be effected in order to achieve the required sensitivity. Furthermore, there are still a surprising number of occasions when, even with the benefit of MS-MS, interferences are present which co-chromatograph to the detriment of the method. In both instances, some form of sample pretreatment is still essential for precise and accurate quantification.

Whilst the future is always difficult to predict, it seems clear that new solid phase extraction phases, based on immunoextraction and possibly molecular imprinted polymers, will be developed to improve specificity of extraction, and that the automation of extraction procedures will proceed apace. However, it seems likely that, for the foreseeable future, standard techniques such as liquid-liquid extraction and conventional solid phase extraction will continue to play a large part in the sample preparation procedures used in bioanalysis.

Bibliography

Blau, K. and Halket, J. (eds.) (1994) *Handbook of Derivatives for Chromatography*, J. Wiley, Chichster, UK.

Chamberlain, J. (ed.) (1995) *Analysis of Drugs in Biological Fluids*, CRC Press, Boca Raton, USA.

Pawliszyn, J. (1997) *Solid-Phase Microextraction: Theory and Practice*, Wiley-VCH, New York, USA.

Reid, E. (ed.) (1976) *Assay of Drugs and other Trace Compounds in Biological Fluids*, North-Holland, Amsterdam, The Netherlands.

Reid, E. (ed.) (1978) *Blood Drugs and other Analytical Challenges*, Ellis Horwood, Chichester, UK.

Reid, E. (ed.) (1980) *Trace Organic Sample Handling*, Ellis Horwood, Chichester, UK.

Reid, E. and Leppard, J.P. (eds.) (1982) *Drug and Metabolite Isolation and Identification*, Plenum, New York, USA.

Reid, E. and Wilson, I.D. (eds.) (1984) *Drug Determination in Therapeutic and Forensic Contexts*, Plenum, New York, USA.

Reid, E., Scales, B. and Wilson, I.D. (eds.) (1986) *Bioactive Analytes, Including CNS Drugs, Peptides, and Enantiomers*, Plenum, New York, USA.

Reid, E., Robinson, J.D. and Wilson, I.D. (eds.) (1988) *Bioanalysis of Drugs and Metabolites, Especially Anti-inflammatory and Cardiovascular*, Plenum, New York, USA.

Reid, E. and Wilson, I.D. (eds.) (1990) *Analysis for Drugs and Metabolites Including Anti-Infective Agents*, Royal Society of Chemistry, Letchworth, UK.

Reid, E. and Wilson, I.D. (eds.) (1992) *Bioanalytical Approaches for Drugs, Including Anti-Asthmatics and Metabolites*, Royal Society of Chemistry, Letchworth, UK.

Reid, E., Hill, H.M. and Wilson, I.D. (eds.) (1994) *Biofluid and Tissue Analysis for Drugs Including Hypolipidaemics*, Royal Society of Chemistry, Letchworth, UK.

Reid, E., Hill, H.M. and Wilson, I.D. (eds.) (1996) *Biofluid Assay for Peptide-Related and other Drugs*, Royal Society of Chemistry, Letchworth, UK.

Stevenson, D. and Wilson, I.D. (eds.) (1994) *Sample Preparation for Biomedical and Environmental Analysis*, Plenum, New York, USA.

Thurman, E.M. and Mills, M.S. (1998) *Solid-Phase Extraction: Principles and Practice*, Wiley-Interscience, New York, USA.

References

Ayrton, J., Dear, G.J., Leavens, W.J., Mallett, D.N. and Plumb, R.S. (1998) Use of generic fast gradient liquid chromatography-tandem mass spectroscopy in quantitative bioanalysis. *J. Chromatogr. B*, **709** 243-54.

Battu, C., Marquet, P., Fauconnet, A.L., Lacassie, E. and Lachâtre, G. (1998) Screening procedure for 21 amphetamine-related compounds in urine using solid phase microextraction and gas chromatography-mass spectrometry. *J. Chromatogr. Sci.*, **36** 1-7.

Boos, K-S. and Rudolph, A. (1998a) The use of restricted-access media in HPLC. Part 1. Classification and review. *LC-GC Int.*, **11** 84-95.

Boos, K-S. and Rudolph, A. (1998b) The use of restricted-access media in HPLC. Part 2. Applications. *LC-GC Int.*, **11** 224-33.

Campanero, M.A., Zamarreño, A.M., Diaz, M., Dios-Viéitez, M-C. and Azanza, J.R. (1997a) Development and validation of an HPLC method for determination of amphotericin B in plasma and sputum involving solid phase extraction. *Chromatographia*, **46** 641-46.

Campanero, M.A., Zamarreño, A.M., Simón, M., Dios, M.C. and Azanza, J.R. (1997b) Simple and rapid determination of piperacillin and ceftazidime in human plasma samples by HPLC. *Chromatographia*, **46** 374-80.

Carlucci, G., Mazzeo, P. and Palumbo, G. (1996) Simultaneous determination of lomefloxacin, fenbufen and felbinac in human plasma using high-performance liquid chromatography. *Chromatographia*, **43** 261-64.

Carlsson, B. and Norlander, B. (1998) Solid phase extraction with end-capped C2 columns for the routine measurement of racemic citalopram and metabolites in plasma by high-performance liquid chromatography. *J. Chromatogr. B*, **702** 234-39.

Cheng, Y-F., Phillips, D.J. and Neue, U. (1997) Simple and rugged SPE method for the determination of tetracycline antibiotics in serum by HPLC using a volatile mobile phase. *Chromatographia*, **44** 187-90.

Cooper, J.D.H., Turnell, D.C., Green, B., Demarais, D. and Rasquin, P. (1994) The analysis of primary and secondary free amino acids in biological fluids: a completely automated process using on-line membrane sample preparation, precolumn derivatisation with o-phthalaldehyde, 9-fluorenyl methyl chloroformate and high-performance liquid chromatographic separations. In *Sample Preparation for Biomedical and Environmental Analysis* (eds. D. Stevenson and I.D. Wilson) Plenum, New York, USA, pp. 87-116.

Escoriaza, J., Dios-Viéitez, M.C., Trocóniz, I.F., Renedo, M.J. and Fos, D.F. (1997) Quantitative analysis of temazepam in plasma by capillary gas chromatography: application to pharmacokinetic studies in rats. *Chromatographia*, **44** 169-71.

Gleixner, A., Sauerwein, H. and Meyer, H.H.D. (1997) Detection of the anabolic steroid methyltestosterone in hair by HPLC-EIA. *Chromatographia*, **45** 49-51.

Heintz, L., Österlind, E., Alkner, U. and Marko-Varga, G. (1997) An integrated coupled column liquid chromatography system for the determination of leukotriene metabolites in biological samples. *Chromatographia*, **46** 365-73.

Jönsson, B.A. and Åkesson, B. (1997) Analysis of 5-hydroxy-N-methyl-2-pyrrolidone and 2-hydroxy-N-methylsuccinimide in plasma. *Chromatographia*, **46** 141-50.

Koide, I., Noguchi, O., Okada, K., Yokoyama, A., Oda, H., Yamamoto, S. and Kataoka, H. (1998) Determination of amphetamine and methamphetamine in human hair by headspace solid phase microextraction and gas chromatography with nitrogen-phosphorus detection. *J. Chromatogr. B.*, **707** 99-104.

Kulczykowska, E. and Iuvone, P.M. (1998) Highly sensitive and specific assay of plasma melatonin using high-performance liquid chromatography with fluorescence detection preceded by solid phase extraction. *J. Chromatogr. Sci.*, **36** 175-78.

Kumazawa, T., Sato, K., Seno, H., Ishii, A. and Suzuki, O. (1996) Extraction of local anaesthetics from human blood by direct immersion-solid phase microextraction (SPME). *Chromatographia*, **43** 59-62.

Lee, H.S., Kim, K., Kim, J.H., Do, K.S. and Lee, S.K. (1997) Simultaneous determination of parathion and metabolites in serum by HPLC with column-switching. *Chromatographia*, **44** 473-76.

Lindenblatt, H., Krämer, E., Holzmann-Erens, P., Gouzoulis-Mayfrank, E. and Kovar, K.-A. (1998) Quantitation of psilocin in human plasma by high-performance liquid chromatography and electrochemical detection: comparison of liquid-liquid extraction with automated on-line solid phase extraction. *J. Chromatogr. B*, **709** 255-63.

Maickel, R.P. (1984) Separation science applied to analyses on biological samples. In *Drug Determination in Therapeutic and Forensic Contexts* (eds. E. Reid and I.D. Wilson), Plenum, New York, USA, pp. 3-16.

Majors, R.E. (1993) A comparative study of European and American trends in sample preparation. *LC-GC Int.*, **10** 93-101.

Martin, P., Morgan, E.D. and Wilson, I.D. (1995) An investigation of the properties of a 'shielded' phase for the solid phase extraction of acidic and basic compounds. *Anal. Proc.*, **32** 179-82.

Martin, P., Morgan, E.D. and Wilson, I.D. (1996) Comparison of the properties of a normal and base deactivated bonded silica gel for the solid phase extraction of [^{14}C]-propranolol. *J. Pharm. Biomed. Anal.*, **14** 419-27.

Martin-Esteban, A., Kwasowski, P. and Stevenson, D. (1997) Immunoaffinity-based extraction of phenylurea herbicides using mixed antibodies against isoproturan and chlortoluron. *Chromatographia*, **45** 364-68.

Mayes, A.G. and Mosbach, K. (1997) Molecularly imprinted polymers: useful materials for analytical chemistry? *Trends Anal. Chem.*, **6** 321-32.

Negrusz, A., Wacek, B.C., Toerne, T. and Bryant, J. (1997) Quantitation of verapamil and norverapamil in postmortem and clinical samples using liquid-liquid extraction, solid phase extraction and HPLC. *Chromatographia*, **46** 191-96.

Paterson, C.J., Boughtflower, R.J., Higton, D. and Palmer, E. (1997) An investigation into the application of capillary electrochromatography-mass spectrometry (CEC-MS) for the analysis and quantification of a potential drug candidate in extracted plasma. *Chromatographia*, **46** 599-604.

Rashid, B.A., Kwasowski, P. and Stevenson, D. (1996) The development of a single-step solid phase extraction for morphine. *Pharm. Sci.*, **2** 115-16.

Rashid, B.A., Aherne, G.W., Katmeh, M.F., Kwasowski, P. and Stevenson, D. (1998) Determination of morphine in urine by solid phase immunoextraction and high-performance liquid chromatography with electrochemical detection. *J. Chromatogr. A*, **797** 245-50.

Rashid, B.A., Briggs, R.J., Hay, J.N. and Stevenson, D. (1997) Preliminary evaluation of a molecular imprinted polymer for solid phase extraction of tamoxifen. *Anal. Commun.*, **34** 303-305.

Roberts, D.W., Ruane, R.J. and Wilson, I.D. (1989) The use of graphitized carbon black in solid phase extraction: comparison with C18-bonded silica gel. *J. Pharm. Biomed. Anal.*, **7** 1077-86.

Ruane, R.J. and Wilson, I.D. (1987) The use of C18-bonded silica in the solid-phase extraction of basic drugs: possible role for ionic interactions with residual silanols. *J. Pharm. Biomed. Anal.*, **5** 723-27.

Sellegren, B. (1997) Noncovalent molecular imprinting: antibody-like molecular recognition in polymeric network materials. *Trends Anal. Chem.*, **16** 310-20.

Shahtaheri, S.J., Kwasowski, P. and Stevenson, D. (1998) Highly selective antibody-mediated extraction of isoproturon from complex matrices. *Chromatographia*, **47** 453-56.

Stevenson, D., Rashid, B.A., Katmeh, M.F. and Kwasowski, P. (1998) Illustrative immunoaffinity procedures. In *Drug Development Assay Approaches Including Molecular Imprinting and Biomarkers* (eds. E. Reid, H.M. Hill and I.D. Wilson), Royal Society of Chemistry, Cambridge, UK, pp. 52-54.

Tan, S.C., Jackson, S.H.D., Swift, C.G. and Hutt, A.J. (1997) Enantiospecific analysis of ibuprofen by high-performance liquid chromatography: determination of free and total drug enantiomer concentrations in serum and urine. *Chromatographia*, **46** 23-32.

Tashkov, W. (1996) Determination of formaldehyde in foods, biological media and technological materials by headspace gas chromatography. *Chromatographia*, **43** 625-27.

Toreson, H. and Eriksson, B.-M. (1997) Liquid chromatographic determination of fluvastatin and its enantiomers in blood plasma by automated solid phase extraction. *Chromatographia*, **45** 29-34.

Watanabe, T., Namera, A., Yashiki, M., Iwasaki, Y. and Kojima, T. (1998) Simple analysis of local anaesthetics in human blood using headspace solid phase microextraction and gas chromatography-mass spectrometry-electron impact ionization selected ion monitoring. *J. Chromatogr. B*, **709** 225-32.

Wulff, W. (1995) Molecular imprinting in cross-linked materials with the aid of molecular templates: a way towards artificial antibodies. *Angew. Chem. Int. Ed. Engl.*, **34** 1812-32.

Yu, G.Y.F., Murby, E.J. and Wells, R.J. (1998) Gas chromatographic determination of residues of thyreostatic drugs in bovine muscle tissue using combined resin mediated methylation and extraction. *J. Chromatogr. B*, **703** 159-66.

Yu, Z. and Westerlund, D. (1997) Influence of mobile phase conditions on the clean-up effect of restricted-access media precolumns for plasma samples injected in a column-switching system. *Chromatographia*, **44** 589-94.

9 Polymers and polymer additives
H.J. Vandenburg and A.A. Clifford

9.1 Introduction

Plastics are complex mixtures of polymers and lower molecular weight components, such as monomers, low oligomers, additives, processing aids and contaminants from feedstocks and processing. The content of these lower molecular weight components needs to be determined for several reasons. Additives, such as antioxidants, ultraviolet absorbers and plasticisers, are essential for the correct functioning of the plastic in its intended use. The manufacturer and processor, therefore, need to be able to determine that the correct levels are present. Plastics are widely used for food packaging and processing, and components can migrate out of the plastic into the food. The level of resulting food contamination is controlled by legislation (EEC, 1989, 1990 and 1992), and manufacturers and regulatory authorities need to be able to determine the levels of such components in the plastic.

As a plastic usually contains several components, analysis of the levels whilst still in the plastic are not usually possible. Some additives can be analysed without extraction by nuclear magnetic resonance spectrometry (Schilling and Kuck, 1991), UV spectrometry (Brauer *et al.*, 1995), and UV desorption/mass spectrometry (Wright *et al.*, 1996), but the complexity of most plastics precludes this. The additive is usually completely extracted from the plastic before analysis. This can be achieved either by dissolution of the polymer and all other components followed by reprecipitation of the polymer, or liquid/solid extraction.

9.2 Dissolution/reprecipitation

The polymer is dissolved in a solvent. The polymer usually needs to be removed from solution before analysis. This is often done by adding a second liquid in which the polymer is less soluble. The polymer then precipitates, leaving the low molecular weight components in solution. There is no possibility of low molecular weight material remaining trapped in the polymer after dissolution; however, inclusion of material in the precipitated polymer is a possibility. A British Standard method (BS, 1965) describes the dissolution of polymers in refluxing toluene, with reprecipitation of the polymer by addition of ethanol. The dissolution process typically takes 30 min to 1 h. After precipitation of the polymer, the solution is filtered and the clear filtrate can be analysed using various techniques. Some polymers are soluble in few solvents, e.g. poly(ethylene terephthalate) (PET), which has been dissolved in hexafluoro-2-propanol/dichloromethane (DCM) mixtures, and the polymer precipitated by addition of acetone or methanol (Barns *et al.*, 1995; Komolprasert *et al.*, 1995).

The addition of a second solvent is not always necessary. Decalin, heated to 110°C, has been used as a solvent for poly(ethylene) (PE) (Schabron and Fenska, 1980) and, heated to 150°C, for poly(propylene) (PP) (Freitag, 1983). The high molecular weight polymer precipitates out as the solution is cooled, and the supernatant solution is filtered and analysed. Reverse-phase (RP) chromatography is the most common method for additive analysis, and organic solvents must often be removed before analysis, which is difficult in the case of solvents with a high boiling point, such as decalin. More volatile solvents can be used if the dissolution is carried out under pressure in an autoclave. Heptane was used to dissolve polyolefins at 160°C (Macko *et al.*, 1995 and 1996); as the polymer is not soluble in heptane at room temperature it precipitates on cooling. The heptane can be readily removed and the sample redissolved in reverse phase liquid chromatography solvents.

After the precipitation step, there is often a considerable amount of oligomeric material ('waxes') in solution, which may need to be removed before further analysis. Some authors have considered this too time-consuming (Spell and Eddy, 1960). Thermoset cross-linked polymers will not dissolve in any solvent and, therefore, cannot be analysed by this method. Therefore, liquid/solid extractions are often the method of choice.

9.3 Liquid/solid extraction

The analyte is extracted from the solid medium by a liquid, and is then separated by physical means, such as filtration. In order for the analyte to dissolve in the liquid and be removed for analysis, it must reach the surface of the polymer. This is usually the rate-limiting step of extractions from polymers. The transfer to the surface occurs by a process of diffusion and, therefore, increasing the rate of diffusion and decreasing the distance the analyte has to diffuse are of key importance in reducing the time taken for extraction. A reduction in the distance is achieved by decreasing the particle size, usually by grinding. However, the heat generated during grinding can cause loss of volatile components or degradation of heat sensitive materials. Therefore, grinding is usually performed under liquid nitrogen to cool the polymer. For spherical particles, the rate of extraction is inversely proportional to the square of the radius, so that grinding a 3 mm pellet to 0.5 mm particles should reduce the extraction time by a factor of 36. The time for complete extraction will be that for the largest particles, and it is this rather than the average particle size which is important.

The diffusion rate can be increased by increasing the temperature. The diffusion coefficient is related to the temperature by the Arrhenius equation:

$$D = D_0 \exp(-E/RT) \qquad (9.1)$$

where, D_0 is a constant, E the activation energy, R the gas constant, and T the absolute temperature. Diffusion coefficients for additive diffusion through polymers are typically of the order of 10^{-10} cm^2/s^1 at 40°C and activation energies of about 100 kJ/mol^1 (Kumar and Prausnitz, 1975). Therefore, increasing the temperature from 20 to 100°C would increase the diffusion coefficient over 30,000 times. Swelling the polymer with solvent also has a significant effect on the diffusion rate. Diffusion of ethylbenzene through polystyrene swollen with CO_2 has been reported to be 10^6 times faster than through the unswollen polymer (Dooley et al., 1995).

Spell and Eddy (1960) studied the extraction of additives from polypropylene at room temperature, and found a large variation in extraction time for different solvents and additives. After powdering the polymer to 50 mesh size, 98% extraction of butylated hydroxytoluene (BHT) was achieved by shaking with carbon disulphide at room temperature for 30 min. To achieve the same recovery with iso-octane required 125 min, and a period of 2000 min was required to recover Santonox with iso-octane. However, solvents which swell the polymer are more likely to dissolve oligomers and produce extracts which are less clean. Optimum extractions are, therefore, achieved by using finely-ground particles with a swelling, but non-dissolving solvent, at as high a temperature as possible.

Extraction techniques can be divided into 'traditional' and 'new'. The traditional techniques include Soxhlet extraction, boiling under reflux and sonication. All these methods are used at atmospheric pressure. The newer methods of supercritical fluid extraction (SFE), accelerated solvent extraction (ASE; a trademark of the Dionex corporation. In this chapter, the initials ASE will be used generically to describe all high pressure liquid extraction methods, whether on Dionex equipment or not), and microwave-assisted extraction (MAE) can all be used at pressures above atmospheric. These techniques are discussed below.

9.4 Traditional methods of solid/liquid extraction

9.4.1 Soxhlet extraction

Soxhlet extraction is a widely-used method, although one of the slowest. The advantages of this process are that the polymer is constantly extracted with condensed solvent containing no extracted analyte and, therefore, total extraction is theoretically possible without equilibrium being set up between polymer and solvent. Once the extraction is started, there is little 'hands on' attention required. However, the extractions can take a very long time, typically from 6 h (Crompton, 1968) to 48 h (Majors, 1970; Haney and Dark, 1980). Even with long extraction times, recovery is not always good. Perlstein (1983) obtained recoveries of only 59% for extraction of Tinuvin 320 from unground polyvinyl chloride (PVC) after 16 h Soxhlet extraction with diethyl ether. However, recoveries rose to 97% from ground polymer. The choice of solvent is significant for the duration of the extraction. Wims and Swarin (1975) found that talc-filled PP needed 72 h extraction with chloroform, but only 24 h with tetrahydrofuran (THF). Thus, small particles are often essential to complete the extraction in a reasonable time, and the solvents must be carefully selected to swell the polymer.

The long duration of the extractions is largely due to the low temperatures at which the extraction takes place. The extraction solvent is the reflux condensate and is much cooler than the boiling liquid. The use of modern, automated equipment (Soxtec), in which the sample is initially immersed in the boiling liquid before being raised out for a final Soxhlet-style extraction should increase the extraction rates significantly.

9.4.2 Boiling under reflux

The sample is added to solvent, which is boiled under reflux. The solvent is, therefore, at the highest temperature possible without applying an external pressure. For this reason, the extractions tend to be much faster than Soxhlet extractions. Whilst, in theory, partition of the analyte between the polymer and solvent prevents complete extraction, this seldom seems to be a problem in practice. The quantity of solvent is much larger than that of polymer, and the partition coefficients usually favour the solvent, resulting in very low levels in the polymer at equilibrium. Any solvent or solvent mixture can be used. Some authors report the use of mixtures to exploit different properties of each, e.g. cyclohexane/2-propanol mixtures for extraction of antioxidants from polyolefins (Nielson, 1993). The cyclohexane is to swell the polymer, whilst the 2-propanol enhances the solubility of the antioxidant.

Caceres and co-workers (1996) compared several methods for extraction of Tinuvin 770 (bis[2,2,6,6-tetramethylpiperidine-4-yl] sebacate) and Chimassorb 944 (poly[N-1,1,3,3-tetramethylbutyl]-N',N''-di[2,2,6,6-tetramethylpiperidinyl]-N',N''-melamino-di-trimethylene) from high density polyethylene (HDPE) pellets. Room temperature diffusion into chloroform and ultrasonication gave less than 20% extraction. Soxtec extraction with DCM for 4h resulted in only 50% extraction. Dissolution of the polymer in dichlorobenzene at 160°C for 1h followed by reprecipitation of the polymer with 2-propanol gave 65–70% recovery. The most successful method was boiling under reflux with toluene at 160°C for 2–4h, which extracted 95% of both additives. The relatively poor performance of the Soxtec extraction compared to the reflux extraction is probably due to the large difference in temperature between the boiling solvents. The pellets were not ground and the size was not specified.

The advantages of this method are that the equipment is very simple and the extractions are faster than Soxhlet extraction. The disadvantage is that, in theory, complete extraction can never be attained and solvents which give rapid extractions often also dissolve lower oligomers.

9.4.3 Sonication

The use of ultrasonic baths has been reported to enhance the extraction rate from polymers. Brandt (1961) extracted tris-(nonylphenol)phosphite (TNPP) from a styrene-butadiene polymer, using 2×20 min extractions with iso-octane as solvent. This compares with 2×1 h extractions for boiling under reflux. Nielson (1991) used ultrasonic extraction of a variety of analytes from PP, low- and high-density polyethylene (LDPE and HDPE). For all samples, the ultrasonic extraction could be achieved within 1h, providing the samples were stirred every 10 min. Further experiments by Nielson (1993) on extraction from HDPE using the same regimen confirmed these results, but extractions were not compared to those without sonication.

Caceres and co-workers (1996) used the same solvent mixtures as Nielson (1991) to extract Tinuvin 770 and Chimassorb 944 from HDPE. The additives could only be extracted at less than 20% recoveries from pellets using ultrasonic extractions of up to 5h. The size of the pellets is not given, but the fact that the sample was not ground may be the reason for the difference in results. Extraction of Chimassorb 81 (2-hydroxy-4-n-octoxybenzophenone) from LDPE and ethylene-vinyl acetate polymer (EVA) was achieved with 6h standing of the sample under DCM (maceration) followed by 3×20 min sonication (Nerin et al., 1996). The initial maceration time allowed swelling of the polymer. The extraction time using maceration alone was 48 h. However, Vandenburg and co-workers (1999b) found that the extraction rate of Irganox 1010 (pentaerythrityl-tetrakis(3,5-di-tert-butyl-4-hydroxyphenyl)-propionate) from PP using either sonication or shaking the sample at the same temperature were the same. Overall, ultrasonic extraction from polymers has given some reasonably fast extractions, but advantages over shaking the sample have not been widely demonstrated.

9.5 High pressure solid/liquid extraction methods

9.5.1 Supercritical fluid extraction (SFE)

A supercritical fluid has properties between those of a gas and a liquid. Thus, the diffusivity is much higher than in liquids, allowing rapid penetration of the sample, and the density is much higher than a gas, which gives the fluid reasonable solvating power. The most common solvent used in SFE is CO_2 on line, which is characterised as a nonpolar solvent with a solubility parameter similar to hexane. It does, however, have some affinity with slightly polar molecules because of its molecular quadrupole. Polar modifiers are sometimes added to the solvent to increase the polarity of the supercritical phase and the solubility of polar compounds. Because the solvating power is lower than that of liquids, the extractions may be limited since not all of the analyte at the surface of the polymer dissolves in the fluid. In this case, the extraction is solubility- rather than diffusion-limited. There is an added complication, because raising the temperature to increase diffusion rates lowers the density of the CO_2 and therefore the solubility. Thus, increasing the temperature in a solubility-limited extraction will slow down the extraction rate. However, the solvent is continually flushed from the cell during dynamic extractions, and so material diffusing to the surface of the polymer will eventually be dissolved in fresh solvent. The combined effect of high temperatures, high pressures, swelling of the polymer by CO_2 and dynamic extraction often gives rise to much faster extractions than Soxhlet or reflux methods.

Bartle and co-workers (1990) described a simple model for extraction from spherical particles, called the 'hot ball' model. In this, it was assumed that the only limiting step was transfer out of the matrix by a process which could be modelled as diffusion. This successfully predicted the characteristics of the extraction curve of $\ln(m/m_0)$ versus time for extraction of BHT from PP (where m_0 is the initial concentration in the plastic and m is the quantity remaining in the plastic). These characteristics are that the curve falls steeply initially, as the analyte is extracted from the surface, and then becomes linear (Figure 9.1). There

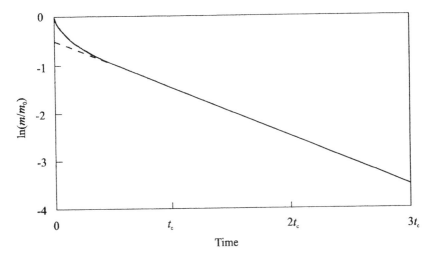

Figure 9.1 Theoretical curve of $\ln(m/m_0)$ versus time for the 'hot ball' model. Abbreviations: m_0, the initial concentration in the plastic; m, quantity remaining in the plastic.

were deviations from the predicted onset of the linear portion and of the extrapolated intercept of the linear portion with the m/m_0 axis. These were explained in terms of nonspherical particles, solubility limitations and nonuniform distribution of the analyte in the polymer. The success of the model indicates that the basic processes can be modelled as diffusion. However, the deviations indicate that solubility and other factors are also significant.

The linear part of the logarithmic plot can be extrapolated to determine the concentration of an extractant without complete extraction, by using three extraction periods of equal duration. The quantity m_0 can be found using the formula:

$$m_0 = m_1 + ([m_2]^2/[m_2 - m_3]) \qquad (9.2)$$

The first extraction must be long enough, such that the second and third extractions are on the linear portion of the curve. The extrapolation method was successful in giving the amount of BHT extractable from PP in three 15 min extractions. This method can eliminate the need for grinding the sample. The model was extended to cover polymer films and nonuniform distribution of the extractant using extraction of cyclic trimer in PET as an example (Bartle et al., 1991). The effect of solubility on extraction was incorporated into the model by Bartle and co-workers (1992). The effects of pressure and flow rate on extraction have been explained theoretically by Clifford and co-workers (1995). Two parameters were defined, firstly, a diffusion term, the diffusion coefficient of the analyte in the polymer divided by the square of the sphere radius (D/a^2), and, secondly, a parameter which is proportional to solubility (ha). The (m/m_0) plot for different values of ha is shown in Figure 9.2. In each case, the curve falls onto a linear portion at longer times.

The concentration of analyte in the polymer during extraction is shown in Figure 9.3. Initially in (a), it is assumed to be constant across the sphere. As material is extracted from the surface of the sphere, the profile moves to (b). This corresponds to the nonlinear portion of the curves in Figure 9.2. Once this profile is established, it reduces in size but maintains the same shape, as shown in (c), during the final exponential decay. If ha (solubility) is large, the vertical portion of the profiles in (b) and (c) are very small and the nonexponential part of the extraction curve is more important. As ha approaches infinity, the model becomes the simple 'hot ball' model, limited only by diffusion. If ha is small, the curved portion of the profiles in (b) and (c) are very flat and nearly the whole extraction

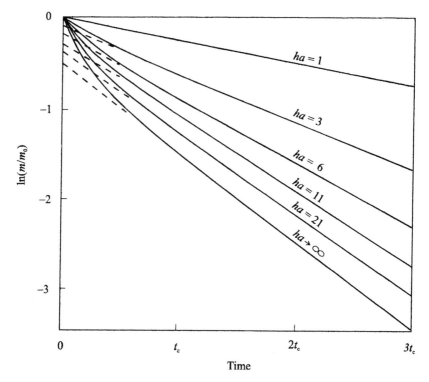

Figure 9.2 ln (m/m_0) versus time for different values of ha (solubility). Abbreviations: m_0, the initial concentration in the polymer; m, quantity remaining in the polymer; ha, a parameter which is proportional to solubility.

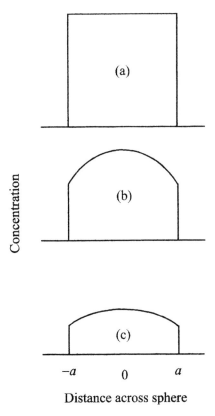

Figure 9.3 Concentration profile of additive in a polymer during extraction. Initially in (A), the concentration is assumed to be constant across the sphere. As material is extracted from the surface of the sphere, the profile moves to (B). Once this profile is established, it reduces in size but maintains the same shape, as shown in (C), during the final exponential decay.

curve is exponential. This corresponds to a completely solubility-limited extraction.

If the pressure is varied at constant flow rate and temperature, both parameters D/a^2 and ha are found to change. Thus, the recovery curves must be fitted for individual pressures, and this has been done for the extraction of Irgafos 168 (tris-(2,4-di-tert-butyl) phosphite) from PP at various pressures (Clifford et al., 1995) (Figure 9.4). The particles were irregular spheres of 0.8±0.2 mm diameter and extraction was carried out at 45°C with pure CO_2 at a flow rate of 7 ml/s, measured with a bubble flowmeter at 20°C and 1 bar (Bartle et al., 1991). The parameters obtained from the fitting are given in Table 9.1.

The values of ha have the form of an isotherm of solubility versus pressure for substances in a supercritical fluid (Figure 9.5). The values of

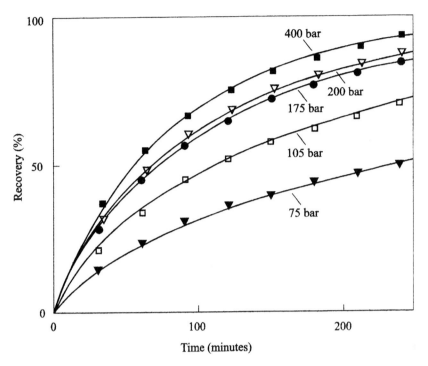

Figure 9.4 Extraction curves at different pressures for extraction of Irgafos 168 from polypropylene (PP). Points = experimental data, solid lines = fitted curve.

Table 9.1 Values of the parameters obtained by fitting the data shown in Figure 9.5

Pressure (bar)	$D/a^2 \times 10^5$ (s^{-1})	ha
75	21	3.2
105	48	5.8
175	90	7.3
200	100	8.1
400	160	8.2

Abbreviations: D/a^2, the diffusion coefficient of the analyte in the polymer divided by the square of the sphere radius; ha, a parameter which is proportional to solubility.

D/a^2 also rise with pressure, and this is explained by higher absorption of the supercritical fluid at higher pressures, causing the polymer to swell and raising the diffusion coefficient. Thus, with polymers, increasing the pressure can be beneficial to SFE, even above pressures where the solubility is no longer rising.

This model predicts that extractions should, therefore, be carried out at as high a temperature and pressure as possible. However, the polymer will

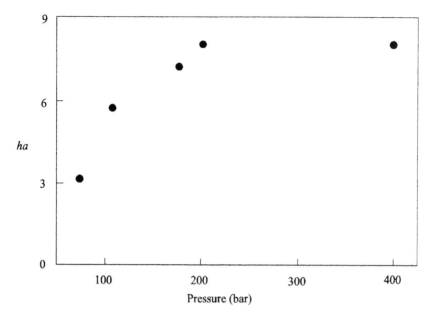

Figure 9.5 Values of *ha* versus pressure for extraction of Irgafos 168 from polypropylene (PP). Abbreviation: *ha*, a parameter which is proportional to solubility.

eventually soften and melt, and the particles will coalesce into one lump. This increases the distance the analyte needs to diffuse and therefore slows extraction. The absorption and swelling of the polymer by CO_2 also reduces the melting and glass transition temperatures, so the maximum temperature cannot be set at the usual melting point of the polymer. Amorphous or rubbery polymers are most able to absorb CO_2 and swell, and it is in these cases that increasing the pressure is most likely to induce melting of the polymer at a fixed temperature. Burgess and Jackson (1992) found that extraction of carbon tetrachloride from chlorinated poly(isoprene) could be completed within 40 min at 60°C and 21 MPa, and not at higher temperatures or pressures. The normal T_g of poly(isoprene) is 120°C, but the softening point is lowered by the CO_2 at high pressures. Therefore, increasing the pressure at high temperatures lowered the softening point and the polymer particles coalesced, reducing the surface area. In this case, at 40°C the extraction improves with increasing pressure. At this low temperature, the extraction is unlikely to be solubility-limited and, therefore, the greater extraction is probably due to increased swelling of the polymer at higher pressures.

The effect of solubility limits becoming important at higher temperatures is illustrated in Figure 9.6 (Lou *et al.*, 1996), which shows the amount of monomer, dimer and trimer extracted from nylon in a fixed

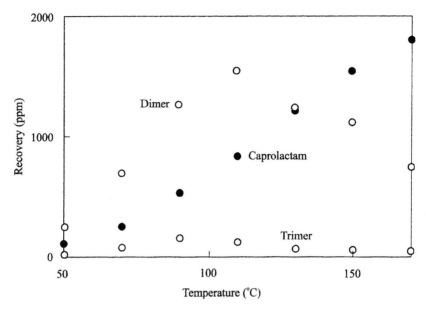

Figure 9.6 Extraction of monomer, dimer and trimer from nylon at different temperatures by supercritical fluid extraction (SFE) with pure CO_2. (Data from Lou et al., 1996)

time and pressure at increasing temperatures. Here, the extraction efficiency increases with temperature until a maximum is reached. This maximum occurs at a lower temperature for the higher molecular weight analytes. This is because of solubility limitations. When solubility becomes the limiting factor, increasing temperature reduces solubility in the fluid and so extraction rates drop. The solubility is lower for higher molecular weight oligomers, therefore the solubility limit is reached at a lower temperature.

Addition of a modifier which swells the polymer also increases extraction rates, both by swelling the polymer and increasing the solubility of the analytes. This can be seen in Figure 9.7 (Lou et al., 1996), where addition of methanol accelerates extraction from nylon. Addition of benzene and hexane, which do not interact significantly with nylon, had very little effect. However, addition of benzene during a static extraction of anti-oxidant from PE powder (Lou et al., 1995) improved recoveries. Benzene was selected as it is known to swell PE, and this was thought to be the main reason for the greater recoveries.

SFE is, therefore, an effective method of extraction from polymers. There are several factors which can affect the success of SFE: temperature; pressure; time; addition of a modifier; the matrix; and the

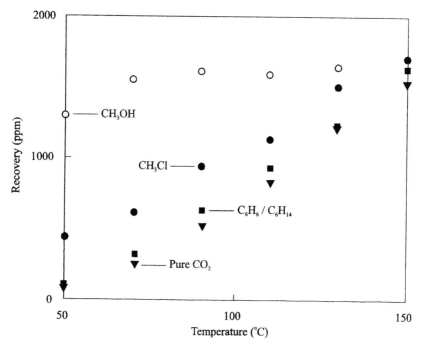

Figure 9.7 Effect of addition of different modifiers to CO_2 in extraction of caprolactam from nylon. (Data taken from Lou *et al.*, 1995).

compound extracted. The interaction of these variables is particularly complex for extraction from polymers, partly because the solvent can interact with the polymer. A rule of thumb for optimum extraction conditions is for dynamic extraction, with the temperature just below the softening point of the polymer under experimental conditions, the pressure as high as possible, and with about 10% addition of a swelling solvent as modifier.

9.5.2 *Accelerated solvent extraction (ASE)*

ASE is a relatively new technique and few reports have been published on its use in extraction from polymers. The sample is loaded into a cell, solvent is pumped in under pressure and it is heated to temperatures of up to 200°C. A period of static extraction follows, and the solvent is then pumped into a collecting vial. A similar volume of solvent is flushed through the cell to transfer all the extracted material to the collecting vial. This has the advantage of being able to use any solvent at high temperatures to achieve the benefits of high dissolving power and swelling of the polymer. However, in practice, the selection of the solvent is not straightforward. Solvents which swell the polymer in Soxhlet and reflux methods will dissolve or soften the polymer at temperatures slightly above their boiling point. Softening causes the agglomeration problems seen with SFE, and dissolved polymer reprecipitates in the transfer lines of the instrument, causing blockages. Therefore, the advantages of faster diffusion gained at high temperatures cannot be achieved with swelling solvents. However, solvents which do not interact at all with the polymer will not swell the polymer even at temperatures at which the polymer will melt or the additive may decompose. Therefore, the solvent must be selected to swell, but not dissolve, the polymer significantly at high temperatures.

Lou and co-workers (1997) used hexane in extractions from nylon, even though it gives poor recoveries during Soxhlet extraction. They pointed out that selection of a suitable extraction solvent is probably the most difficult step in optimising ASE, as there are few data on the solubility of polymers in solvents at high temperatures. They investigated the effect of temperature, pressure and flow rate on extraction from nylon and poly(butylene terephthalate) (PBT) using a homemade ASE. Pressure and flow rate had little effect on extraction rates, which would be expected if the extraction is not solubility-limited. The amount extracted increased with increasing temperature. This is also shown in Figure 9.8 for extraction of Irganox 1010 from PP (Vandenburg et al., 1998). Increasing temperature increases diffusion rates and the polymer-solvent interaction, allowing greater swelling. Therefore, extraction rates increase with temperature.

The same polymers had previously been analysed using SFE with pure and modified CO_2 (Lou et al., 1996) and the result using pure CO_2 at 170°C and 30.7 MPa compared. The recoveries using ASE for caprolactam from nylon and the dimer and trimer from PBT were 1.1, 6.5 and 37.6 times higher, respectively, than those obtained with SFE. However, in these conditions the SFE was not optimum, particularly for the dimer and trimer, where the peak extraction after 30 min occurred at 110 and 90°C, respectively (as shown in Figure 9.6) (Lou et al., 1996). This extraction peak at low temperatures clearly indicates solubility-limited extractions. Addition of a modifier (methanol for nylon and chloroform for PBT) during the static extraction stage further increased recoveries from SFE, particularly for the dimer and trimer, but recoveries with ASE were still higher by approximately 1.5 times for dimer and trimer. In this case, the modifier was only added during the initial static extraction, and no experiments were performed with modified CO_2 during the dynamic extraction, but this would probably have enhanced the SFE even more.

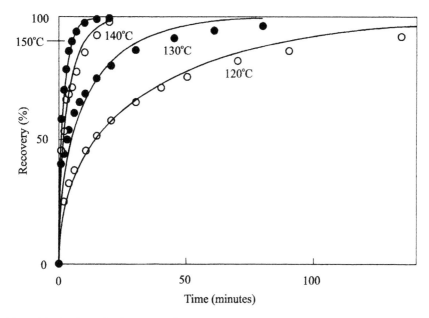

Figure 9.8 Accelerated solvent extraction (ASE) of Irganox 1010 from polypropylene (PP) with 2-propanol at different temperatures. Points = experimental data, solid lines = curve fitted using the 'hot ball' model. (From Vandenburg et al., 1998).

From these results, it appears that ASE offers significant advantages over SFE with CO_2 alone for extraction of compounds with a low solubility in CO_2. However, addition of modifiers to the CO_2 can result in a similar performance for both techniques.

The solubility of additives is expected to be high in the solvents at elevated temperatures, and therefore solubility-limiting behaviour should be rare in ASE. This was found to be the case for extraction of Irganox 1010 from PP (Vandenburg et al., 1998). Extractions were analysed using the hot ball model developed for SFE, assuming no solubility limits (i.e. ha was infinite). Figure 9.8 shows that the extraction data fitted the model quite well. Extractions in 2-propanol, acetone and cyclohexane were carried out over a 40°C temperature rise, up to the point at which the polymer softened. The diffusion coefficients were calculated from the hot ball model, and increased by a factor of 22, 12 and 6 for extractions with 2-propanol, acetone and cyclohexane, respectively. Therefore, heating the swelling solvent (cyclohexane) had much less effect than heating the non-swelling solvent (2-propanol), because heating the non-swelling solvent caused the polymer and solvent to interact, and so the polymer swelled at high temperatures. For the swelling solvent, the polymer was already swollen, and hence heating had much less effect.

The problem in selecting a solvent for an extraction is how to determine which will swell the polymer at high temperatures. A possible scheme involves the Hildebrand solubility parameter (Vandenburg et al., 1998). The Hildebrand Parameter, δ, is defined as the square root of the internal energy of vaporisation divided by the molar volume, also called the cohesive energy density. Broadly speaking, a polymer will be more soluble in a solvent if the solubility parameters are similar. The greater the difference in solubility parameter between the solvent and polymer, the greater the temperature at which the polymer will become soluble. Also, at a fixed temperature, a solvent with a closer solubility parameter will swell the polymer more, and thus give faster extractions. Therefore, for extractions from PP at 120°C, the extraction rate increases with decreasing difference in solubility parameter (Vandenburg et al., 1998). Acetone (difference in $\delta = 3.7\,\text{MPa}^{1/2}$) gives the fastest extractions, followed by 2-propanol (difference in $\delta = 6.9\,\text{MPa}^{1/2}$) with acetonitrile the slowest (difference in $\delta = 7.7\,\text{MPa}^{1/2}$). If the solubility parameter was too close, as is the case with chloroform, the polymer softens and dissolves.

However, the predictions are qualitative, and it is still difficult to select the most appropriate solvent. One approach to overcoming this difficulty is to select a solvent with a solubility parameter several units ($\text{MPa}^{1/2}$) different from the polymer. This is the non-swelling solvent. Extractions are carried out using this solvent to determine the maximum temperature at which the polymer will soften or dissolve. If the polymer is not softened at a temperature at which the analyte may decompose, then a second solvent with a solubility parameter close to the polymer is added incrementally until the polymer dissolves. In this way, maximum swelling can be achieved at the optimum extraction temperature. Hildebrand parameters for several common solvents and polymers are presented in Table 9.2.

Test extractions showed that this method of solvent selection could be used for a variety of polymers (Vandenburg et al., 1999a). For nylon, the optimum conditions were hexane/ethanol (60:40) at 170°C, and extractions from ground material were 95% complete after 16 min, including warm-up time. For extraction of plasticiser from PVC, optimum conditions were hexane/ethyl acetate (60:40) at 170°C, giving effectively complete extractions after 13 min. Irganox 1010 could be extracted from PP in 18 min, using 2-propanol/cyclohexane (97.5:2.5).

ASE can be used to give very fast extractions. The main problem is solvent selection, but Hildebrand solubility parameters can be used to optimise solvent selection with a few simple experiments.

Table 9.2 Solubility parameters of common solvents and polymers

Material	Solubility parameter ($MPa^{1/2}$)	Reference
Hexane	14.9	Barton, 1983
Cyclohexane	16.8	Barton, 1983
Ethyl acetate	18.6	Barton, 1983
Dichloromethane	19.8	Barton, 1983
Acetone	20.3	Barton, 1983
2-Propanol	23.5	Barton, 1983
Acetonitrile	24.3	Barton, 1983
Ethanol	26.0	Barton, 1983
Methanol	29.7	Barton, 1983
Polypropylene	16.6	Barton, 1990
PVC	19.5	Barton, 1990
PET	20.5	Barton, 1990
Nylon 66	28	Barton, 1990
PMMA	19.0	Barton, 1990

Abbreviations: PVC, polyvinyl chloride; PET, polyethylene terephthalate; PMMA, poly(methyl methacrylate).

9.5.3 Microwave-assisted extraction

Sample and solvent are placed in a container and heated using microwave energy. The technique has evolved from closed-vessel microwave acid digestions. The apparatus typically consists of closed vessels with temperature and pressure control, allowing the solvent to be heated under pressure above its normal boiling point and remain liquid. The solvent must contain a component with a high dielectric constant to be heated by microwaves. Carousels of extraction vessels may be used, allowing for simultaneous extraction of up to 12 samples. In many ways, the advantages are similar to those found in ASE, i.e. that the benefits of high temperatures can be used with low boiling solvents through the use of pressure. The difference in heating mechanism has some effect. If the solvent has a high dielectric constant, then the heating will be faster than the use of an oven. However, if the solvent has a low dielectric constant, heating will be slow or nonexistent. Similar considerations for solvent selection apply here as for ASE, with the exception that dissolution of the polymer will not block any transfer lines. Complete dissolution of the sample is likely to lead to rapid extractions, the polymer precipitating when the solvent cools. However, partial dissolution and softening of the polymer will result in agglomeration of particles and a reduction in extraction rate.

Although the thermal effects are significant, there is a possibility of nonthermal effects of microwave radiation on the extraction. Most polymers are effectively transparent to microwaves, and therefore polymer/microwave interactions are unlikely. The microwaves could, in principle, interact with the analyte and enhance the extraction. Jickells and co-workers (1992) attempted to determine whether there were any nonthermal effects in migration from microwave cookware into food simulants (olive oil and solvents). From the study of a number of polymers and migrants, the only effect observed was due to the heating effect of the microwaves.

So far, there have been few reported uses of MAE for extraction from polymers. Freitag and John (1990) used an acetone/heptane (1:1) mixture to extract additives from LDPE, HDPE and PP. After grinding to 20 mesh, 91–97% of Irgafos 168, Chimassorb 81 and Irganox 1010 were extracted within 6 min from HDPE and within 3 min from PP and LDPE. Dissolution of larger particles in toluene/1,2-dichlorobenzene was effected in 5 min. However, the dissolution method gave only 85% recovery of the additives from the PE samples, although 95% recovery was achieved from the PP. Extractions from larger pellets were less efficient. The vessels were pressurised, but the temperatures and pressures reached were not reported.

Nielson (1991) compared microwave extraction with sonication for HDPE, PP and LDPE and a variety of additives. BHT, Irganox 1010 and Irganox 1076 could be extracted at >90% recoveries from ground HDPE in 20 min at 50% power. Two solvent systems were used, 2-propanol/cyclohexane (1:1) and DCM/2-propanol (98:2). In each case, the 2-propanol was present to absorb the microwave energy and the other solvent to swell the polymer. The sample was stirred at 5 min intervals. Butylated hydroxyethylbenzene (BHEB), Isonox 129 (2,2'-ethylidene-bis(4,6-di-tertiarybutylphenol)) and erucamide (13-*cis*-docosenamide) slip agent were extracted from LDPE using the same conditions. A variety of additives could be extracted from ground PP in 20 min at 20% power, with stirring required every 5 min. Only Irganox 3114 (1,3,5-tris[3,5-di-*tert*-butyl-4-hydroxybenzyl]s-triazine 2,4,6[1H, 3H, 5H] trione) had a low recovery of 79%. When larger pellets were used, the recoveries were also high, except for Irganox 1010, for which only 50% recovery was possible without grinding. The vessels were not pressurised and the temperature using the DCM/2-propanol mixture did not exceed 50°C.

Costley and co-workers (1997) reported the extraction of cyclic trimer from PET using a variety of solvents heated to 120°C in pressurised vessels. The polymer fused at temperatures above 120°C with DCM, therefore higher temperatures were not investigated. Two hour extraction with DCM at 120°C gave the same extraction as 24 h Soxhlet extraction with xylene as solvent. MAE using hexane-acetone (1:1), water, acetone and acetone-DCM (1:1) all gave much lower recoveries than DCM at

120°C. This would be expected from the solubility parameters (Table 9.2). At the same temperature, DCM would swell the polymer most. Vandenburg *et al.* (1999b) compared extractions of Irganox 1010 from PP with ASE, MAE and conventional techniques. In this case, the preferred solvent was 2-propanol, therefore microwave heating was efficient. Extraction using MAE gave the same recovery as exhaustive refluxing with chloroform with 5 min extraction at 150°C and 3 min warm-up time. However, the sample extract could not be removed from the cell until it was cool, which took 20 min. ASE and MAE both gave significantly faster extractions than any conventional technique with equal recoveries.

From these examples, MAE appears to be a rapid and effective technique for polymer extractions. The solvent can be selected to swell the polymer, provided some microwave absorbing solvent is present. The polymer needs to be ground for efficient extraction.

9.6 Comparison of techniques

As is often the case, there is no clear method which is preferable in all situations. Conventional techniques of dissolution/precipitation, Soxhlet and reflux extraction are well documented and use simple, inexpensive equipment. However, they cannot be automated, the extraction times are long and, in the case of dissolution methods, often require subsequent clean-up steps. Therefore, these methods are probably most suitable for laboratories with only occasional requirements for polymer analysis, with boiling under reflux the best 'all round' performer. With outlay of significant funds, automated equipment can be used (Soxtec), which combines the best qualities of reflux and Soxhlet extractions. However, little evidence has been reported concerning the efficacy of this system for polymer analysis.

The use of high pressure offers the fastest, automated extraction methods. SFE, ASE and MAE all significantly improve on extraction rates achieved with conventional methods but require the purchase of special equipment. Again, choosing between these techniques is not simple. The greatest body of published evidence concerns the efficacy of SFE. However, optimisation of the extractions is not simple. Many factors have to be considered, in particular choice of temperature, pressure and modifier, and therefore method optimisation can be time-consuming. Sample sizes tend to be much smaller than in conventional methods, which raises the possibility of unrepresentative samples being used. The use of the extrapolation procedure allows the determination of analyte concentration without total extraction, which may remove the need to grind samples.

ASE offers the greatest flexibility of solvents and solvent mixtures. Up to 24 consecutive analyses can be programmed independently. Some initial experiments are required to optimise the method to give rapid extractions without dissolving the polymer, but Hildebrand solubility parameters can be a useful aid in this. ASE would be ideal for a laboratory which analysed a large number of different polymers.

MAE can extract up to 12 samples simultaneously, if the solvents are the same in each cell. The solvents used must contain a component which absorbs microwaves strongly. This makes the selection of solvents more problematical than for ASE, as for example pure hydrocarbons cannot be used. Extraction times are similar to ASE but, as 12 samples can be analysed together, the average extraction time per sample can be much less for replicate extractions. Therefore, MAE may be the preferred method for a laboratory analysing large numbers of similar samples.

Each technique, therefore, offers some advantages and disadvantages, the final choice is dependent on the requirements of the user and the particular application.

References

Barns, K.A., Damant, A.P., Startin, J.R. and Castle, L. (1995) Qualitative liquid-chromatographic atmospheric-pressure chemical-ionization mass-spectrometric analysis of polyethylene terephthalate ligomers. *J. Chromatogr. A.*, **712** 191-99.

Bartle, K.D., Boddington, T., Clifford, A.A. and Hawthorne, S.B. (1992) The effect of solubility on the kinetics of dynamic supercritical fluid extraction. *J. Supercrit. Fluids*, **5** 207-12.

Bartle, K.D., Boddington, T., Clifford, A.A., Cotton, N.J. and Dowle, C.J. (1991) Supercritical fluid extraction and chromatography for the determination of oligomers in PET films. *Anal. Chem.*, **63** 2371-77.

Bartle, K.D., Clifford, A.A., Hawthorne, S.B., Lagenfield, J.J., Miller, D.J. and Robinson, R. (1990) *J. Supercrit. Fluids*, **3** 143-149.

Barton, A.F.M. (1990) *Handbook of Polymer-Liquid Interation Parameters and Solubility Parameters*, CRC Press, Boca Raton.

Barton, A.F.M. (1983) *Handbook of Solubility Paramerters and Other Cohesion Parameters*, CRC Press, Boca Raton.

Brandt, H.J. (1961) Determination of tris(nonylated phenyl) phosphite in styrene-butadiene synthetic polymer by a direct UV spectrophotometric method. *Anal. Chem*, **33** 1390-91.

Brauer, B., Funke, T. and Schulenbergschell, H. (1995) Determination of stabilizer concentration in polyethylene. *Deutsche Lebens. Rundsch.*, **91** 381-85.

British Standard 2782 (1965) Part 4, Method 405D.

Burgess, A.N. and Jackson, K. (1992) The removal of carbon-tetrachloride from chlorinated polyisoprene using carbon-dioxide. *J. Appl. Polym. Sci.*, **46** 1395-99.

Caceres, A., Ysambert, F., Lopez, J. and Marquez, N. (1996). Analysis of photostabiliser in high density polyethylene by reverse- and normal-phase HPLC. *Sep. Sci. Technol.*, **31** 2287-98.

Clifford, A.A., Bartle, K.D. and Zhu, S.A. (1995) Supercritical-fluid extraction of polymers—theorectical explanation of pressure and flow-rate effects. *Anal. Proc.*, **32** 227-30.

Costley, C.T., Dean, J.R., Newton, I. and Carroll, J. (1997) Extraction of oligomers from poly(ethylene terephythalate) by microwave-assisted extraction. *Anal. Comm.*, **34** 89-91.

Crompton, T.R. (1968) Identification of additives in polyolefins and polystyrenes. *Europ. Polym. J.*, **4** 473-96.

Dooley, K.M., Launey, D., Becnel, J.M. and Caines, T. (1995) Measurement and modeling of supercritical-fluid extraction from polymeric matrices. *ACS Symposium Series*, **608** 269-80.

EEC 89/109 (1989) Council Directive 89/109/EEC of 21 December 1988 on the approximation of laws of the Member States relating to materials and articles intended to come into contact with foodstuffs. *Off. J. Europ. Comm.*, **L40** 38-44.

EEC 90/128 (1990) Corrigendum to Commission Directive 90/128/EEC of 23 February 1990 relating to materials and articles intended to come into contact with foodstuffs. *Off. J. Europ. Comm.*, **L349** 26-47.

EEC 92/39 (1992) Commission Directive 92/39/EEC of 14 May 1992 amending Directive 90/128/ EEC relating to materials and articles intended to come into contact with foodstuffs. *Off. J. Europ. Comm.*, **L168** 21-29.

Freitag, W. and John, O. (1990) Fast separation of stabilizers from polyolefins by microwave-heating. *Angew. Makromol. Chem.*, **175** 181-85.

Freitag, W. (1983) Determination of a polymeric light stabilizer (Chimassorb-944) in polypropylene, *Fres. Z. Anal. Chem.*, **316** 495.

Haney, M.A. and Dark, W.A. (1980) *J. Chromatogr. Sci.*, **18** 655-59.

Jickells, S.M., Gramshaw, J.W., Castle, L. and Gilbert, J. (1992) The effect of microwave-energy on specific migration from food contact plastics. *Food Addit. Contam.*, **9** 19-27.

Komolprasert, V., Lawson, A.R. and Hargreaves, W.A. (1995) Analytical method for quantifying butyric-acid, alathion, and diazinon in recycled poly(ethylene-terephthalate). *J. Agric. Food Chem.*, **43**, 1963-65.

Kumar, R. and Prausnitz, J.M. (1975) Solvents in chemical technology, in *Solutions and Solubilities* (ed. M.R. Dack), John Wiley and Sons, New York.

Lou, X.W., Janssen, H.G. and Cramers, C.A. (1996) Effects of modifier addition and temperature-variation in SFE of polymeric materials. *J. Chromatogr. Sci.*, **34** 282-90.

Lou, X.W., Janssen, H.G. and Cramers, C.A. (1995) Investigation of parameters affecting the supercritical-fluid extraction of polymer additives from polyethylene. *J. Microcol. Sep.*, **7** 303-17.

Lou, X.W., Janssen, H.G. and Cramers, C.A. (1997) Parameters affecting the accelerated solvent extraction of polymeric samples. *Anal. Chem.*, **69** 1598-603.

Macko, T., Furtner, B. and Lederer, K. (1996) Analysis of aromatic antioxidants and ultraviolet stabilisers in polyethylene using high temperature extraction with low boiling solvent. *J. Appl. Polym. Sci.*, **62** 2201-207.

Macko, T., Siegl, R. and Lederer, K. (1995) Determination of phenolic antioxidants in polyethylene with dissolution by N-heptane in an autoclave. *Angew. Makromol. Chem.*, **227** 179-91.

Majors, R.E. (1970) High speed liquid chromatography of antioxidands and plasticizers using solid core supports. *J. Chromatog. Sci.*, **8** 339.

Nerin, C., Salagranca, J. and Cacho, J. (1996) Separation of polymer and online determination of several antioxidants and UV stabilizers by coupling size-exclusion and normal-phase high-performance liquid-chromatography columns. *Food Addit. Contam.*, **13** 243-50.

Nielson, R.C. (1991) Extraction and quantitation of polyolefin additives. *J. Liq. Chromatogr.*, **14** 503-19.

Nielson, R.C. (1993) Recent advances in polyolefin additive analysis. *J. Liq. Chromatogr.*, **16** 1625-38.

Perlstein, P. (1983) The determination of light stabilizers in plastics by high-performance liquid-chromatography. *Anal. Chim. Acta*, **149** 21-27.

Schabron, J.F. and Fenska, L.E. (1980) Determination of BHT, Irganox 1076, and Irganox 1010 antioxidant additives in polyethylene by high performance liquid chromatography. *Anal. Chem.*, **52** 1411-15.

Schilling, F.C. and Kuck, V.J. (1991) Determination of stabilizer concentrations in polyethylene. *Polym. Degradat. Stabil.*, **31** 141-52.

Spell, H.L. and Eddy, R.D. (1960) Determination of additives in polyethylene by absorption spectroscopy. *Anal. Chem.*, **32** 1811.

Vandenburg, H.J., Carlson, R., Clifford, A.A., Bartle, K.D., Carroll, J. and Newton, I.D. (1999a) A simple solvent selection method for accelerated solvent extraction of additives from polymers. *Anal. Chem.* (submitted).

Vandenburg, H.J., Clifford, A.A., Bartle, K.D., Carroll, J. and Newton, I.D. (1999b) Comparison of accelerated solvent extraction (ASE) microwave assisted extraction of additives from polypropylene. *Analyst* (accepted).

Vandenburg, H.J., Clifford, A.A., Bartle, K.D., Carroll, J., Newton, I.D. and Zhu, S. (1998) Factors affecting high pressure solvent extraction (accelerated solvent extraction) of additives from polymers. *Anal. Chem.*, **70** 1943-48.

Wims, A.M. and Swarin, S.J. (1975) Determination of antioxidants in polypropylene by liquid chromatography. *J. Appl. Polym. Sci.*, **19** 1243-56.

Wright, S.J., Dale, M.J., Langridge-Smith, P.R.R., Zhan, Q. and Zenobi, R. (1996) Selective *in situ* detection of polymer additives using laser mass spectrometry. *Anal. Chem.*, **68** 3585-94.

10 Environmental applications

Pat Sandra, Frank David, Erik Baltussen and Tom De Smaele

10.1 Introduction

Growing concern about the quality and safety of our environment has led regulatory bodies (such as the European Union and the Environmental Protection Agency) to compile lists of priority pollutants and to establish rules and regulations for their control. As a consequence, an array of analytical procedures has been developed over the years to determine a wide variety of micropollutants in air, in drinking-, surface-, ground-, rain- and waste-water, and in soil, sludge, sediment and solid waste. At present, many laboratories focus on aspects such as cost-effectiveness, high sample throughput, enhanced analyte detectability and the identification potential of unknown pollutants.

In the analysis of environmental samples, enrichment is of vital importance because samples are too dilute or too complex and need to undergo a chain of specific treatments to make them compatible with analytical techniques. While the dictum 'the best sample preparation is no sample preparation' is also true for environmental analysis, direct analysis is only applicable in some exceptional cases, for example, the determination of polycyclic aromatic hydrocarbons (PAHs) in drinking water by liquid chromatography (LC) with fluoresence detection.

An important factor in micropollutant analysis is the amount of solvent used in performing enrichment of the micropollutant traces from the different matrices. State-of-the-art procedures are designed to reduce the consumption of organic solvents, which are often more toxic than the traces of micropollutants to be determined, or, even better, to avoid the use of solvents completely. Using 100 ml dichloromethane (DCM) in liquid-liquid extraction (LLE) of water samples to determine 0.1 µg/l PAHs may serve as a typical example of how not to proceed.

This chapter describes novel methods for extraction of micropollutants from different matrices. Old technologies, such as continuous LLE, Soxhlet extraction, etc., are not discussed. Extraction methods which have not been treated in other chapters of this book are discussed in detail.

10.2 Key extraction methods

10.2.1 Introduction

Extraction can be performed with a gas, a liquid or a solid and in a static or dynamic mode. Different combinations are applied in environmental analysis.

10.2.2 Air and gaseous samples

10.2.2.1 Adsorptive extraction (AE)

Most procedures currently used for the preconcentration of volatile organic compounds (VOCs) in air and gaseous samples are based on adsorption of the analytes of interest onto a suitable adsorbent material (Ventura *et al.*, 1995; Helmig and Greenberg, 1994; Peters and Renesse van Duivenbode, 1994) followed either by liquid or thermal desorption. Thermal desorption is nowadays preferred because it guarantees higher detectability. In the thermal desorption method, the desorbed analytes are refocused in a cold trap prior to transfer onto the analytical column. Common adsorbents include carbon-based materials, such as activated carbon and carbon molecular sieves (Tang *et al.*, 1993; Knobloch and Engewald, 1995), and porous organic polymers, such as Tenax and Chromosorb (Jüttner, 1988). These are all relatively strong adsorbents giving excellent performance for nonpolar (semi)-volatiles (benzene, toluene, ethylbenzene, xylene [BTEX], PAHs, polychlorinated biphenyls [PCBs]). Unfortunately, their application to the analysis of polar solutes is rather limited. Lack of retention during sampling is not generally the problem because polar analytes are strongly retained on most adsorbents. This strong retention, however, often precludes rapid and complete desorption, resulting in low recoveries and a severe risk of carry-over. Moreover, the long residence times of the analytes on the hot and active adsorbent surface during desorption might result in reactions of the analytes initiated by the surface itself or with other adsorbed species. These reactions can result in permanent adsorption and/or in artifact formation, which are clearly undesirable effects (David *et al.*, 1998).

For the enrichment of target compounds, selective adsorptive extraction or adsorption with reaction can be applied. Examples include the determination of aldehydes and ketones in air on a 2,4-dinitrophenylhydrazine impregnated silica gel adsorbent, and of ethylene oxide on a hydrogen bromide (HBr) impregnated adsorbent, forming the hydrazones and 2-bromo-ethanol, respectively. Another complicating factor when working with adsorbents is that the organics which eventually have to be determined can be formed because of degradation of the adsorbent itself. This is, for example, the case with Tenax, where acetophenone and benzaldehyde are often detected, and with Chromosorb leading to styrene and α-methylstyrene. From the above, it is clear that alternatives to the classic adsorbents are necessary for adequate handling of air samples containing polar solutes.

10.2.2.2 Static sorptive extraction
Static sorptive extraction is nowadays very well known under the name solid phase microextraction (SPME) (Pawliszyn, 1997; see also Chapter 4). In our opinion, the latter nomenclature is incorrect, as a sorption rather than an adsorption mechanism occurs on the solid surface. Polymers, such as polydimethylsiloxane (PDMS), polyacrylate (PA), polyethylene glycol (PEG), etc., behave as liquids at room temperature and a partitioning mechanism applies. SPME is not very useful for air monitoring because by its static character, the volume sampled and the distribution constants must be known. The applicability for quantitative analysis of gaseous samples is, therefore, restricted to headspace sampling above liquid or solid samples (see Sections 10.2.3.4 and 10.2.4.5).

10.2.2.3 Dynamic sorptive extraction
Several years ago, a novel method for the preconcentration of organic components from air was developed by Burger and Munro (1986), Bicchi and co-workers (1987, 1988, 1989) and Roeraade and Blomberg (1986). Open tubular trapping (OTT) columns coated with a thick film of a gas chromatographic (GC) stationary phase were used for sample enrichment. In this method, preconcentration occurs by sorption of the analytes into the bulk of the liquid phase, instead of adsorption onto an active adsorbent surface. The most commonly used GC stationary phase, namely 100% PDMS, has the best characteristics to serve as sorbent. PDMS is a gum and the technique is, therefore, also known as gum phase extraction (GPE). Preconcentration by sorption has some clear advantages over adsorption onto an active surface. In the sorption mode, polar solutes desorb fast at low temperatures due to a weak interaction of the analytes with the PDMS material. Moreover, PDMS is much more inert than a standard adsorbent, minimizing the losses of unstable and/or polar analytes. Another advantage of the PDMS phase is that its degradation products can be easily identified via mass spectrometric (MS) detection, as they generate characteristic silicone mass fragments. Peaks originating from the sorbent cannot, therefore, be mistaken as originating from sampled analytes. For practical purposes, PDMS has the following advantages. Since the analytes are retained in the bulk of the material, retention of the solutes on this phase is more reproducible than in the case of adsorbents. For example, a high water content in the gas sample does

not affect retention of the analytes. In addition, batch-to-batch irreproducibility, as is sometimes encountered when working with adsorbents, is absent in the case of PDMS.

Despite several clear advantages of PDMS open tubular traps over classic adsorbents, they have never gained widespread acceptance. This is because they have several limitations. OTTs have only a limited sample capacity because only a small amount of stationary phase is present per unit of trap length. For adequate retention, long traps are necessary (up to several metres in length). Since the air sample has to be sucked through the trap by applying a vacuum to the outlet, OTTs allow only low sampling flow rates (typically in the order of 10 ml/min). Because an air volume of 0.5–5 l is generally required for adequate detection limits, this implies that long sampling times are necessary. Another disadvantage of OTTs is that an additional GC oven is required for thermal desorption of the OTT with efficient refocusing of the enriched analytes. In an attempt to overcome the problems associated with OTTs, Ortner and Rohwer (1996) designed a multi-channel OTT. This short trap contains several channels in parallel, and should tolerate significantly higher flow rates since the pressure drop over the trap is very small. Unfortunately, this device allows flow rates up to only 15 ml/min. Due to the unfavourable geometry of the trap, at higher flow rates the number of plates becomes too low to ensure quantitative trapping. On the instrumental side, the multi-channel OTT has the advantage that it can be desorbed in a standard GC injector. Recently, traps packed with 100% PDMS particles have been developed to concentrate components from air (Baltussen *et al.*, 1997, 1998a, 1998b). These packed traps have the same advantages as open tubular traps with respect to inertness and thermal desorption characteristics but they allow sampling flow rates as high as 2.5 l/min.

The main difference between the PDMS sorbent and standard adsorbents is that with the PDMS phase the analytes are retained in the bulk of a polymeric liquid phase, whereas on an adsorbent the analytes are retained on an active surface. The retention volume (V_r, ml) for a given analyte is defined as the volume of gas phase from which the analytes of interest are retained in the hypothetical case that the trap has an infinite number of plates (N). On an adsorbent, the retention volume will depend on surface-specific parameters, such as the number of active sites/surface area and the total surface area. An additional parameter for adsorbents is the adsorption constant (k_a). In the case of the PDMS material, V_r will be determined primarily by the amount of stationary phase (V_s) and the equilibrium constant (K). A theoretical model, which allows calculation of retention and breakthrough volumes for packed PDMS traps, has been described (Baltussen *et al.*, 1997). It is based on the breakthrough theory described by Lövkvist and Jönsson (1987). The 5%

breakthrough volume is given by:

$$V_b^5 = V_0 \cdot (1+k) \cdot \left(0.9025 + \frac{5.360}{N} + \frac{4.603}{N^2}\right)^{-1/2} \quad (10.1)$$

which can be simplified to ($V_0 \ll KV_s$):

$$V_b^5 = K \cdot V_s \cdot \left(0.9025 + \frac{5.360}{N} + \frac{4.603}{N^2}\right)^{-1/2} \quad (10.2)$$

where, V_b^5 is the 5% breakthrough volume (for definition see below) (ml), V_0 is the trap void volume (ml), K is the equilibrium constant between gas and PDMS phase, and N is the number of theoretical plates.

Equation 10.2 consists of two parts. The first part ($KV_s = V_r$) is of a thermodynamic nature and governs the equilibrium distribution of the analyte between gas and liquid phase. The second part is of a kinetic nature and accounts for the loss of analytes due to the non-infinite plate number of the trap. The breakthrough percentage is defined as the amount of analyte lost relative to the amount of analyte sampled (Baltussen et al., 1997; Lövkvist and Jönsson, 1987). Values for K can, for example, be calculated from retention data (Baltussen et al., 1997), which are available in the literature for a large number of analytes (Sadtler Standard Gas Chromatography Retention Index, 1985). All remaining parameters in Equation 10.2 can be calculated from theory (Baltussen et al., 1997). The performance of PDMS traps for air monitoring will be illustrated in Sections 10.4.1 and 10.4.2 for nonpolar and polar solutes.

10.2.3 Water and liquid samples

10.2.3.1 Gas phase extraction

Gas phase extraction from water samples is known as purge and trap (P & T). Purge and trap is generally used for the enrichment of nonpolar volatile organic compounds prior to GC analysis. An inert gas is bubbled through the water sample, causing the purgeable organics to move from the aqueous phase to the vapour phase. The volatile compounds are then trapped (solid phase extraction) on an adsorbent, such as Tenax or active charcoal. The trap containing the adsorbent is built into a desorption chamber equipped with a heating mechanism, which, when activated, permits the desorption (gas phase extraction) of the trapped compounds. This technique has the distinct merit of providing a clean sample, free from its often very dirty matrix. A purge and trap device can easily be mounted on a GC equipped with an electron capture detection (ECD) and photoionisation detection (PID) in series, or with a MS. This technique is most appropriate for ppb level analysis of low molecular weight, slightly water-soluble, volatile organics with a boiling point below 200°C. A variation of purge and trap is closed-loop stripping analysis (Grob and Zurcher, 1976). This is a combination of gas phase extraction with solid phase extraction in a closed system.

10.2.3.2 Micro-liquid-liquid extraction (μLLE)
At present, the most widely-used method of sample preparation in water analysis is liquid-liquid extraction (LLE). LLE may be carried out manually by shaking the water sample with an organic solvent in a separation funnel or automatically, using a continuous liquid-liquid extractor. Continuous LLE is recommended by the Environmental Protection Agency (EPA) for the enrichment of semi-volatiles (base-neutral/acid extractables). Depending on the extraction conditions used, extracts can contain intermediate to low polarity, weakly volatile pollutants (universal extraction for neutral semi-volatiles) or acid and base compounds (selective extraction) by adjusting the pH. LLE is very time-consuming and uses toxic solvents. The volume of the extract is usually too large for direct injection and, in order to obtain sufficient sensitivity, an additional evaporation-concentration step (Kuderna-Danish) is necessary. Particular care needs to be taken in both the solvent extraction and concentration procedures to avoid contamination of the sample. Moreover, solvent impurities will be concentrated often masking the target solutes.

At the present time, automated μLLE in the vial of an injection autosampler can offer a simple and robust choice for a number of applications. By using highly sensitive and selective detectors, eventually in combination with large volume injection, sub-ppb sensitivities can be reached. The performance of μLLE will be illustrated with the analysis of trihalomethanes in drinking water (see Section 10.4.3).

10.2.3.3 Solid phase extraction (SPE)
This innovative extraction procedure is gaining wide acceptance, being much faster, cheaper and more versatile than most classical techniques. The principle of retention is analogous to high-pressure liquid chromatography (HPLC) and is suitable for low, intermediate and high polarity pollutants, depending on the solid phase used. Large sample volumes can be handled using relatively small amounts of solid phase, which in turn requires only small volumes of solvent for solid phase stripping. This often eliminates the need for an additional evaporation step and thereby considerably reduces the risk of contamination.

Depending on the sample throughput and the compounds to be analysed, SPE may be performed on cartridges and disks. The latter require less solvent for stripping. Recently, the US EPA method 525, which originally stipulated LLE for the enrichment of semi-volatiles from water samples, has been modified to include SPE. Moreover, SPE disk cartridges were introduced, increasing the practicability of SPE even further. Thermal desorption may sometimes replace solvent elution, thus ensuring the highest degree of sample enrichment; its major drawback being the thermal instability of the solid phase materials. GPE extraction can provide an answer here (see Section 10.2.3.5). The performance of SPE in environmental analysis will be illustrated with the determination of triazines in water samples applying large volume injection-capillary gas chromatography (CGC) (see Section 10.4.6), and of pesticides in river water applying supercritical fluid desorption and supercritical fluid chromatographic analysis (see Section 10.4.7).

10.2.3.4 Static sorptive extraction
Solid phase microextraction (SPME), as outlined in Chapter 4, is a very powerful solventless method of sample preparation used for the determination of organic micropollutants in water samples. Depending on the nature of the target compounds and on the cleanness of the sample, headspace or liquid sampling can be considered. The drawbacks of SPME are: that the method is restricted to target compound analysis; that in order to have high detectability highly sensitive and selective detectors should be used: and, moreover, that quantitation-calibration can be very time-consuming. The performance of SPME will be illustrated with the analysis of pesticides and phenols in water samples via liquid sampling (see Sections 10.4.4 and 10.4.5).

10.2.3.5 Dynamic sorptive extraction
Dynamic sorptive extraction or gum phase extraction (GPE) for water samples is a technique that resembles SPE to some extent. The most important difference is that in GPE retention occurs by dissolution into the bulk of the extractant phase. The retaining power of the PDMS phase for a certain solute is, therefore, not dependent on its concentration or on the concentration of matrix compounds, i.e. humic or fulvic substances, as is the case in adsorbent-based SPE procedures. One hundred per cent PDMS is most successfully used as the retaining phase, though in a recent publication the use of an acrylate polymeric sorbent was described for the enrichment of aliphatic amines (Baltussen *et al.*, 1998c). As described in Section 10.2.2.3, thermal desorption can be used to rapidly transfer the entire sample onto the GC column.

In Section 10.2.2.3, a model was developed for the calculation of retention and breakthrough volumes of solutes on packed PDMS traps for air sampling. In air sampling, k can be calculated directly from gas chromatographic retention indices. When sampling water, this is no longer possible. Fortunately, Pawliszyn and co-workers (Arthur et al., 1992; Potter and Pawliszyn, 1994) found that in SPME there is a good correlation between log $K_{PDMS/w}$ and log $K_{o/w}$, where $K_{PDMS/w}$ and $K_{o/w}$ are the PDMS-water and octanol-water partitioning coefficients, respectively. Therefore, log $K_{o/w}$ values were used to estimate k by:

$$k = \frac{K_{PDMS/w}}{\beta} = \frac{K_{o/w}}{\beta} \qquad (10.3)$$

where, β is the phase ratio of the PDMS trap. Plate numbers were calculated using the Knox equation (Kennedy and Knox, 1972; Guiochon, 1980).

$$h_r = 3v^{1/3} + \frac{1.5}{v} + 0.05v \qquad (10.4)$$

with:

$$h_r = \frac{H}{d_p} = \frac{L}{N \cdot d_p} \qquad (10.5)$$

$$v = \frac{u \cdot d_p}{D_m} \qquad (10.6)$$

After choosing the sampling velocity (u, m/s) and trap parameters (d_p and L, m), the diffusivity of the analyte (D_m, m^2/s) in the mobile phase has to be obtained from published data sets (Lewis, 1995), or calculated using approximation techniques (Wilke and Chang, 1955). N can then be calculated directly from Equations 10.4–10.6. Log $K_{o/w}$ values for many compounds can be found in the literature (Noble, 1993; Verschueren, 1996) and this allows calculation of retention and breakthrough volumes on the PDMS traps. GPE enrichment for phenols from water samples will be discussed in Section 10.4.5.

10.2.4 Soil, sludge, sediment and solid samples

10.2.4.1 Gas phase extraction
Thermal desorption (TD) on-line coupled to CGC-MS is a powerful technique for the analysis of air volatiles collected on adsorbent and sorbent tubes, for the analysis of residual monomers in polymers, for the

determination of residual solvents in pharmaceutical products, etc. The same principle can also be applied to the determination of volatiles in solid environmental samples. The solid material is placed in a TD tube and installed in the TD unit. The temperature is increased typically to 350°C and the volatiles released are transported with the carrier gas stream to a cryotrap. After cryofocusing, the trap is heated and the solutes are introduced into the column with a narrow injection band. The method has been criticized because of the small sample size loaded into the TD cartridge, which is typically in the order of 500 mg. Standard deviations can therefore be quite high but, nevertheless, this simple approach rapidly yields a picture of the pollution. An off-line TD unit for gram quantities recently became available from Gerstel (Mülheim a/d Ruhr, Germany). The on-line approach will be illustrated in Section 10.4.8.

10.2.4.2 Supercritical fluid extraction (SFE)
The principles of SFE have been discussed in Chapter 5. For environmental analysis, the strong point of SFE is the 'selectivity' that can be introduced in modern instrumentation (Medvedovici et al., 1997). A schematic diagram of a 'state-of-the-art' SFE instrument is presented in Figure 10.1.

The system consists of an extraction cartridge (1), an oven (2), a carbon dioxide pumping module (3), an independent modifier pump (4), a variable restrictor or nozzle (5), a solid phase trap (6), and a pump to deliver the rinse solvents (7). A personal computer provides instrumental control. The most interesting feature of SFE, namely its 'selectivity', has not been fully exploited until now in the preparation of environmental samples. By optimizing SFE conditions, it is possible to prepare extracts which are directly amenable to chromatographic analysis. The system shown in Figure 10.1 contains several points at which selectivity can be introduced. In the first instance, the extraction selectivity and efficiency can be controlled by the nature of the supercritical medium (Selectivity 1). The reasons for adding a polar or a nonpolar modifier to the CO_2, normally used as the supercritical medium, are threefold: (i) to increase the solubility; (ii) to destroy the matrix effects; and (iii) to enhance diffusion by swelling of the matrix. Modifiers can also be added to retain unwanted solutes but this counteracts (ii) and (iii). The effect of the addition of a polar modifier on the recovery of PCBs has been described previously (Hawthorne et al., 1992).

The second opportunity for selectivity concerns the density of the supercritical medium and the temperature (Selectivity 2) (David et al., 1993). The importance of the latter parameter is often neglected but, as solubility is also controlled by vapour pressure, this can be exploited to

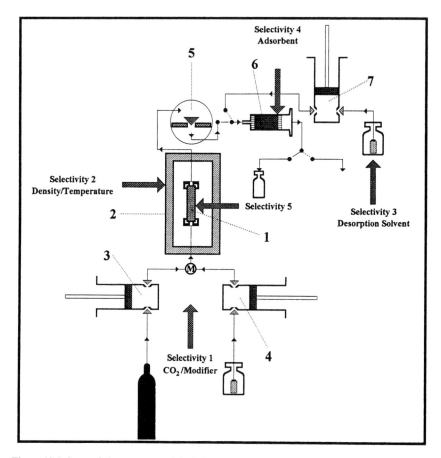

Figure 10.1 State-of-the-art supercritical fluid extraction (SFE) instrument. 1, Extraction cartridge; 2, oven; 3, CO_2 pumping module; 4, independent modifier pump; 5, variable restrictor or nozzle; 6, solid phase trap; and 7, pump to deliver the rinse solvents.

introduce selectivity. This will be illustrated in Section 10.4.10 with the determination of PCBs in lipid matrices. After leaching of the sample, the extract is collected on a solid trap filled with a nonpolar or polar adsorbent, which can be selected according to the application (Selectivity 4). The trap is then rinsed with a solvent, the polarity of which can be chosen to desorb the solutes of interest in a selective way (Selectivity 3). Last but not least, an adsorbent can be added in the extraction thimble (Selectivity 5). This introduces the possibility of extracting liquids by SFE (see Section 10.4.11) but it also facilitates the retention of unwanted polar solutes (fixation) or the enhancement of recoveries of nonpolar solutes (exaltation) (Sandra *et al.*, 1995a).

10.2.4.3 Micro liquid-liquid extraction by ultrasonic treatment (UT)
A method which often provides good results for solid samples is ultrasonic treatment with an organic solvent. Surprisingly, the same quantitative data were obtained for the extraction of PCBs from a certified sediment sample, SM 1939 (BCR, Brussels, Belgium) compared to optimized SFE (David et al., 1992). A sample weighing 0.5 g was extracted with 10 ml hexane in an ultrasonic bath for 30 min and, after filtration and concentration to 0.5 ml, was injected into a CGC-ECD instrument. Nowadays, with large volume injection the concentration step can be omitted.

10.2.4.4 Accelerated solvent extraction (ASE)
ASE is a promising technique for the extraction of a large number of pollutants from solid matrices. The speed of the extraction process is greatly increased compared to conventional methods, e.g. Soxhlet extraction. ASE is applicable to virtually all extractable organics from the priority pollutant lists (Lesnik and Fordham, 1994/1995). Disadvantages of this procedure can be: the lack of selectivity, which means that further clean-up steps are needed; and that the sample is diluted and requires further concentration. By adding alumina to the extraction cell, however, PCBs could be extracted with high selectivity from fish tissue and directly analysed by CGC-ECD (Dionex, 1996). Large volume injection can eventually avoid the concentration step. ASE was evaluated in our laboratory for the extraction of organometallics from a sediment sample. Dibutyltindichloride and tributyltinchloride were quantitatively extracted by applying hexane/acetone in the ratio 1/1 as extracting solvent at 1500 psi and 100°C. After derivatization with methyl magnesiumchloride, the analysis was performed with CGC-atomic emission detection (AED) monitoring the Sn 303 nm emission line.

10.2.4.5 Static sorptive extraction
The performance of SPME for solid samples will be illustrated with the determination of organometallics in sediment and fish tissue (Section 10.4.9).

10.3 Instrumentation

10.3.1 Introduction

In recent years, important new developments have occurred in all separation methods and technological development is proceeding unabated. This has had a direct influence on the development of new methods of sample preparation. Some notable developments will be discussed briefly.

10.3.2 Gas chromatography (GC)

Besides developments in column technology and instrument design, for example, electronic pneumatic control (allowing retention time locking) and fast capillary GC, injection has become more versatile following the introduction of a programmable temperature vaporizer (PTV) inlet (Figure 10.2).

On the one hand, this is a universal injector (allowing hot and cold split/splitless, direct and cool on-column injection) but, on the other hand, due to its solvent venting possibilities, PTV can also serve as an

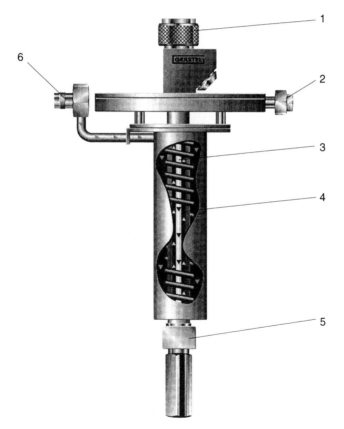

Figure 10.2 Programmable temperature vaporizer (PTV) injector. 1, Septumless sampling head; 2, split vent; 3, glass insert; 4, injection port can be cooled or heated; 5, Graphpack connector; and 6, cryoconnector.

injector for large volumes and as an interface for LC-GC (David *et al*., 1997, 1998). Moreover, the PTV is also a key component in a state-of-the-art thermal desorption unit (Figure 10.3).

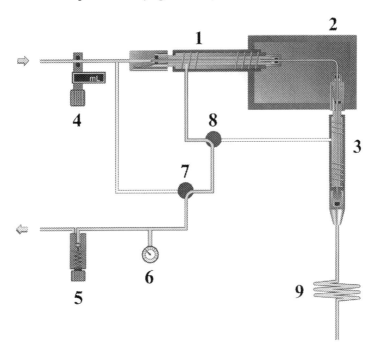

Figure 10.3 Schematic representation of the system applied for thermodesorption. 1, Thermodesorption unit; 2, temperature-controlled transfer capillary; 3, programmable temperature vaporizer (PTV) injector; 4, back-pressure pneumatics with mass-flow controller; 5, back-pressure regulator; 6, pressure gauge; 7, split-splitless valve; 8, 3/2-way solenoid valve; and 9, capillary column.

Environmental laboratories now recognize that reliable data can only be obtained with a mass spectrometer as detector, and there is more to come, such as CGC-time-of-flight mass spectrometry (TOFMS) and CGC-inductively coupled plasma mass spectrometry (ICPMS).

10.3.3 Liquid chromatography (LC)

Microbore columns are now increasingly accepted because of the lower solvent consumption and the increased detectability (Verzele *et al*., 1988). The most important developments are, however, the bench top LC-MS systems operating in the electrospray (ES) and atmospheric pressure chemical ionization (APCI) mode, and the LC-MS-MS systems. Both developments place less stringent requirements on sample preparation.

10.3.4 *Supercritical fluid chromatograhpy (SFC)*

SFC is still alive and will eventually find its way into environmental laboratories. The introduction of new SFC and SFE instrumentation has resulted in exciting new applications (Anton and Berger, 1997).

10.4 Selected applications

10.4.1 *Dynamic sorptive extraction of pollutants in air*

10.4.1.1 *Introduction*

The analysis of PAHs and of nitro-PAHs using sorptive enrichment on packed PDMS traps will be described as an example of the analysis of pollutants in air. PAHs and nitro-PAHs are carcinogenic compounds occurring in air. PAHs originate from human emissions, such as the burning of fossil fuel, whilst nitro-PAHs can be formed in the air by photoreactions. Due to the low concentration at which they occur, classical sampling methods, e.g. sampling on filters, require the use of excessively large air volumes (often more than $1000\,m^3$). As an alternative approach, the air sample can be drawn directly through a thermal desorption tube packed with PDMS. The analytes will dissolve into the PDMS particles and can be released upon heating. Due to the weak sorption strength of PDMS compared to adsorbents, the analytes can be thermally desorbed at mild temperatures. Compared to classical techniques, the thermal desorption method provides superior sensitivity, since the entire sample is transferred to the column. Therefore, only a small sample volume is needed.

10.4.1.2 *Experimental*

Prepacked PDMS traps were obtained from Gerstel (Mülheim a/d Ruhr, Germany). These cartridges contain ca. 400 mg of 100% pure PDMS particles, i.e. no support material or other underlying active surfaces are present. For the determination of the PAHs, only 5 l air samples were used. For the determination of nitro-PAHs, which occur at much lower concentrations, 176 l of air was sampled. Other analytical conditions were

Instrumental configuration:
Gas chromatograph	HP 6890
Inlet	Gerstel Thermodesorption System (TDS-2)
Detector	HP 5972 Mass Selective Detector (MSD)
MSD mode	Selected ion-monitoring (SIM) (2 ions/ component)

Data handling	Mass spectrometry (MS)-Chemstation (Windows Series)
Column	25 m × 0.25 mm ID × 0.25 µm HP-5MS

Experimental conditions:

Desorption temperature	250°C (5 min)
Injection mode	Splitless
Desorption flow	250 ml/min
Carrier gas	Helium
Head pressure	75 kPa
Carrier gas mode	Constant pressure
Oven temperature	40°C (2 min)–15°C/min–325°C

10.4.1.3 Results and discussion

A typical chromatogram obtained from the enrichment of a 5 l urban air sample is presented in Figure 10.4A, together with the analysis of a spiked clean air sample (Figure 10.4B).

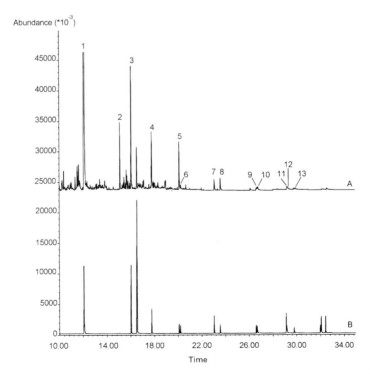

Figure 10.4 Polycyclic aromatic hydrocarbon (PAH) analysis with polydimethylsiloxane (PDMS). A, Actual air sample; B, Standard solutes.

For solute identification and concentration levels see Table 10.1.

Table 10.1 Polycyclic aromatic hydrocarbon (PAH) concentrations in air

No.	Component	Conc. ng/m^3
1	Naphthalene	2035
2	Acenaphthylene	107
3	Acenaphthene	45.3
4	Fluorene	22.2
5	Phenanthrene	47.6
6	Anthracene	51.7
7	Fluoranthene	11.2
8	Pyrene	10.9
9	Benz(a)anthracene	7.68
10	Chrysene	7.56
11	Benz(b)fluoranthene	7.73
12	Benz(k)fluoranthene	7.50
13	Benz(a)pyrene	1.99
14	Indeno(123cd)pyrene	2.73
15	Dibenz(ah)anthracene	1.12
16	Benzo(ghi)perylene	4.19

All EPA priority PAHs were detected in the air sample. The concentration levels were similar to those obtained with classical sampling methods, e.g. collection of particulates on a glass fibre filter followed by Soxhlet extraction. Encouraged by the results obtained for the PAHs, an attempt was made to determine a range of nitro-PAHs in air (Table 10.2).

The chromatogram obtained after preconcentration of a clean air sample spiked at the ng/m^3 level, is presented in Figure 10.5A. All nitro-PAHs are apparent. It was, however, impossible to determine these solutes in a real air sample, because of the complexity of chromatograms obtained in the selected ion-monitoring (SIM) mode. Only 1-nitropyrene

Table 10.2 Nitro-polycyclic aromatic hydrocarbons (PAHs) spiked in clean air

No.	Component
17	1-Nitronaphthalene
18	2-Nitronaphthalene
19	2-Nitrobiphenyl
20	3-Nitrobiphenyl
21	1,5-Dinitronaphthalene
22	1,3-Dinitronaphthalene
23	2,2-Dinitrobiphenyl
24	9-Nitroanthracene
25	1,8-Dinitronaphthalene
26	1-Nitropyrene
27	2,5-Dinitrofluorene

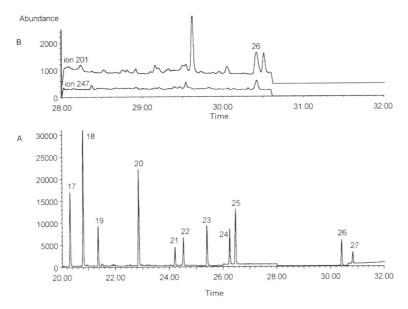

Figure 10.5 Nitropolyaromatic hydrocarbon (nitro PAH) analysis with PDMS. A, Ion-monitoring traces for 3-nitropyrene; B, Standard solutes. (Guiochon, 1980).

could be positively detected by monitoring ion 201 and 247 (Figure 10.5B), and was found to be present at a concentration of 103 pg/m^3. The detectability was in the order of 5 pg/m^3.

10.4.2 Dynamic sorptive extraction of nicotine in hospital air (Baltussen et al., 1998d)

10.4.2.1 Introduction

Cartridges packed with PDMS particles have also been applied to monitor nicotine, the tracer of cigarette smoke, in hospital air before and after filtration. PDMS was found to be far superior in terms of quantitation of nicotine compared to the adsorbents Carbotrap, Tenax and Chromosorb 101. The effectiveness of a nicotine filter based on charcoal is questionable.

10.4.2.2 Experimental

Samples were taken simultaneously in the in- and outlet of the filter. Sample sizes were 1 l, sampled in 10 min at 100 ml/min. The instrumental set-up and conditions were as follows:

Instrumental configuration:
Gas chromatograph   HP 6890
Inlet   Gerstel Thermodesorption System (TDS-2)

Detector	Nitrogen-phosphorus detector (NPD)
Column	15 m × 0.15 mm ID × 2 μm CP-SIL5CB (Chrompack)

Experimental conditions:

Desorption temperature	250°C (5 min)
Injection mode	Splitless
Carrier gas	Helium
Head pressure	145 kPa
Carrier gas mode	Constant pressure
Oven temperature	50°C–10°C/min–200°C

10.4.2.3 Results and discussion

Initial experiments showed that the nitrogen-phosphorus detector (NPD) provided enough sensitivity for the detection of nicotine in ambient air. Two chromatograms, corresponding to the air sample entering and leaving the filter unit, are presented in Figure 10.6A and B, respectively.

It is clear from these figures that the filter withdraws only a certain proportion (30%) of the nicotine.

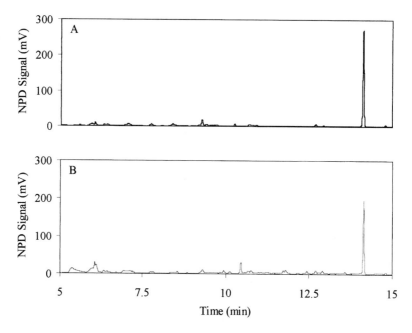

Figure 10.6 Nicotine in hospital air sampled by PDMS. A, Before the filter; B, After the filter. Abbreviation: NPD, nitrogen-phosphorus detector.

10.4.3 Micro-liquid-liquid extraction of trihalomethanes in drinking water

10.4.3.1 Introduction

The trihalomethanes, chloroform, dichlorobromomethane, dibromochloromethane and bromoform, are commonly present in tap water at low concentrations (less than 25 µg/l). These compounds are formed from organic material during chlorination of drinking water. According to World Health Organization (WHO) regulations (1984), the sum of these four compounds should not exceed 100 µg/l. There is, therefore, a need to monitor the presence of these compounds in drinking water using reliable methods. The use of fast analysis and automation saves laboratories time and expense.

Trihalomethanes can be analysed in drinking water samples using various techniques, including static headspace, purge and trap, liquid-liquid extraction and direct aqueous injection (Soniassy *et al.*, 1994; Temmerman *et al.*, 1991). Static headspace and purge and trap are highly sensitive techniques; concentrations below 1 µg/l can easily be detected. However, both techniques require dedicated instrumentation. Although direct aqueous injection gives reliable results and does not require sample preparation, the use of retention gaps that have to be replaced at regular intervals is required. Micro-liquid-liquid extraction with pentane followed by CGC-ECD analysis is a desirable alternative. Micro-liquid-liquid extraction can nowadays be easily performed with automated devices, such as the Hewlett-Packard 7686 PrepStation. The vials containing the two phases are then transferred to an HP 7673 autosampler and automated injection is performed from the upper (pentane) layer.

10.4.3.2 Experimental

Calibration solutions of the four target solutes were prepared in pentane. Water samples were taken directly from a tap. One millilitre aliquots were then pipetted into 2 ml automatic sampler vials (this manual operation could also have been performed by the PrepStation). The remainder of the analytical procedure was automated. The HP 7686 PrepStation was used to dispense 0.5 ml pentane into each sample vial. The compounds were extracted by vortex mixing during 1 min at medium speed on the HP G1296A bar code reader. The two solvent layers were allowed to sit for 1 min to separate. The sample vial was then transferred to the HP 7673 automatic sampler for analysis. The analyses were performed on an HP 6890 GC. Automated splitless injection was carried out using an HP 7673 autosampler with enhanced parameters, which makes it possible to inject from different heights in a sample vial (David *et al.*,

1996). The instrumental configuration and analytical conditions were as follows:

Instrumental configuration:

Gas chromatograph	HP 6890
Inlet	Split/splitless
Detector	Micro ECD
Automatic sampler	HP 7673
Liner	Single taper deactivated
Data handling	Chemstation (DOS Series)
Column	30 m × 0.53 mm ID × 2.65 μmHP-1

Experimental conditions:

Inlet temperature	250°C
Injection volume	1 μl
Injection mode	Splitless
Sampling depth	10.0 mm
Purge time	0.50 min
Purge flow	50 ml/min
Carrier gas	Hydrogen
Head pressure	12 kPa at 35°C
Carrier gas mode	Constant flow
Flow, velocity	5 ml/min, 38 cm/s
Oven temperature	35°C (1 min)–10°C/min–125°C
Detector temperature	340°C
Detector gas	Argon/5%methane at 30 ml/min

10.4.3.3 Results and discussion

A typical chromatogram obtained for a calibration mixture containing 5 ppb of each trihalomethane in pentane is presented in Figure 10.7.

All compounds are well separated within 8 min. In this chromatogram, it appears that the GC run time can be shortened but, when extracted samples are analyzed, the presence of possible interfering peaks indicated the necessity for the longer run time. The linearity of the detector was measured in the range 1–100 ppb, corresponding to the range that is usually monitored in tap water samples. The ECD detector gave a linear response for all four compounds, as shown by the calibration curve for dichlorobromomethane in Figure 10.8. A sample of tap water was analysed five times using these calibration curves (five sample vials, each prepared by the PrepStation and analyzed once with CGC-ECD).

A typical chromatogram for the tap water sample is shown in Figure 10.9. The four trihalomethanes are easily detected. Other unidentified compounds were also present in this tap water sample.

ENVIRONMENTAL APPLICATIONS 263

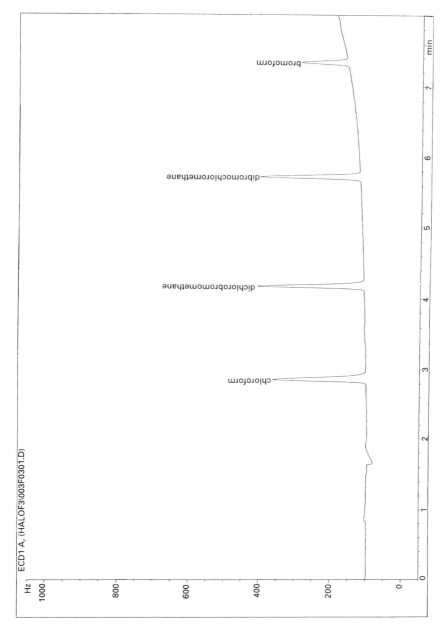

Figure 10.7 Analysis of trihalomethane standards.

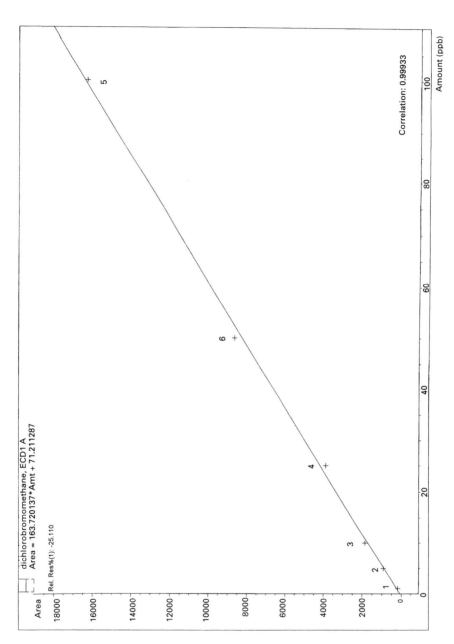

Figure 10.8 Calibration graph for dichlorobromomethane.

ENVIRONMENTAL APPLICATIONS

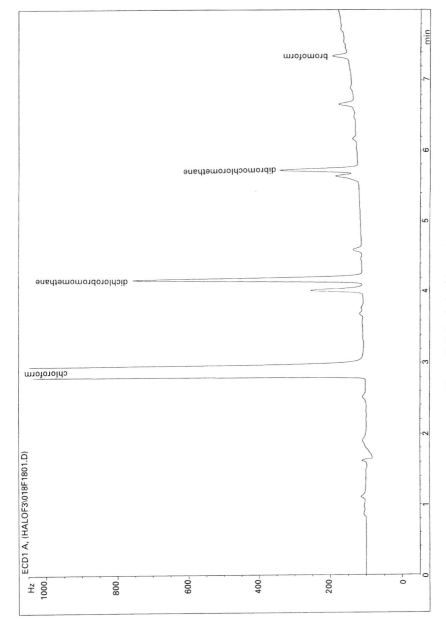

Figure 10.9 Analysis of tap water.

The quantitative data for the five replicate analyses are summarized in Table 10.3.

Table 10.3 Reproducibility of trihalomethane determinations

ppb	$CHCl_3$	$CHBrCl_2$	$CHBr_2Cl$	$CHBr_3$
Mean	4605*	11	3.8	1.1
RSD%	5.2	5.4	6.4	8.7

*: estimated by extrapolation of the curve. Abbreviations: RSD, relative standard deviation.

The concentration of chloroform (an estimation extrapolated from the calibration curve) greatly surpasses the WHO limit. The results of the entire analytical procedure, including pipetting, Prepstation extraction and CGC-ECD analysis, were found to be highly repeatable, with a relative standard deviation (RSD) of < 10% for all compounds. The total sample turn-around time was less than 10 min.

10.4.4 Static sorptive extraction of pesticides from water samples (Sandra et al., 1997)

10.4.4.1 Introduction

Two interlaboratory tests were organized to evaluate the robustness of SPME in terms of accuracy, repeatability and reproducibility, namely the analysis of pesticides in water samples (Gorecki et al., 1996) and of VOCs in water samples (Nilsson et al., 1996). The results were presented at the 18th International Symposium on Capillary Chromatography (Riva de Garda, Italy, May 20–24, 1996). The data on pesticide analysis are presented. Full details on the practical organization, the SPME and instrumental parameters, and the summary of the data can be found in the report by Gorecki and co-workers (1996).

10.4.4.2 Experimental

SPME was performed using a manual SPME device (Supelco, Bellefonte, PA, USA). All extractions were carried out using a 100 μm PDMS fibre. Ten millilitre water samples were spiked with a pesticide mixture, the qualitative composition of which is shown in Table 10.4, resulting in a 50, 30, 10 and 1 ppb concentration level for each pesticide. These samples were used for calibration. An 'unknown' sample was prepared by diluting a spiking solution, containing the pesticides in unknown concentrations, in water. The spiking solution was prepared by the organizers of the SPME interlaboratory test. Prior to use, the fibre was conditioned at 250°C for 2 h by insertion in a split/splitless injector. Extractions were performed in 20 ml headspace vials kept at room temperature. The

Table 10.4 Peak elucidation, retention times, ions and quantitative results

Peak no.	Pesticide	t_R min	Quant. ion	Qual. ion	Found ppb	True value ppb
1	Dichlorvos	13.75	109	185	31	25
2	Eptam	16.25	128	86, 189	11	10
3	Ethoprophos	22.84	158	200, 242	20	17
4	Trifluralin	23.97	306	264	2.5	2
5	Simazine	25.18	201	186, 173	27	25
6	Propazine	25.65	214	58, 229	10	10
7	Diazinon	26.70	304	152, 179	9	10
8	Me-chlorpyriphos	28.58	286	125	2.5	2
9	Heptachlor	28.85	100	272	11	10
10	Aldrin	30.27	263	66, 298	3	2
11	Metolachlor	30.41	162	146, 238	18	17
12	Endrin	35.21	263	209	10	10

Abbreviations: t_R, retention time; Quant., quantification; Qual., qualification.

extraction time was 45 min and the samples were stirred at 1000 rpm during extraction using a magnetic stirrer and a Teflon-coated stir bar. The instrumental configuration and analytical conditions were as follows:

Instrumental configuration:
Gas chromatograph   HP 5890
Inlet   Split/splitless
Detector   HP 5971 MSD–SIM
Liner   0.75 mm ID liner (Supelco)
Data handling   Chemstation (DOS Series)
Column   30 m × 0.25 mm ID × 0.25 µm SPB-5 (Supelco)

Experimental conditions:
Inlet temperature   250°C
Desorption time   5 min
Injection mode   Splitless, valve closed for 5 min
Carrier gas   Helium
Head pressure   0.5 bar
Carrier gas mode   Constant pressure
Oven temperature   40°C (5 min)–30°C · min–100°C–5°C · min–250°C

10.4.4.3 Results and discussion
Detection was effected by ion-monitoring and the ions used for qualitative and quantitative analysis are listed in Table 10.4.

For all pesticides, the calibration curves obtained for concentrations of 1–30 ppb had correlation coefficients greater than 0.995. At higher concentration, saturation of the fibre occurred. As an example, the curves obtained for trifluralin, dichlorvos and aldrin are presented in Figure 10.10. The analysis of the unknown sample is shown in Figure 10.11.

All pesticides are easily detected with excellent peak shape. The quantities were calculated relative to the calibration curves. The results are summarized in Table 10.4. After submission of the results, the true spiked values were received. As can be deduced from Table 10.4, a good correlation was obtained between the measured and the true values. Concerning the sensitivity of SPME, the distribution constant of the solutes between water and PDMS controls the amount sorbed. For highly apolar solutes, such as heptachlor, sensitivities in the order of 10 ppt can be reached. For the more polar dichlorvos, the sensitivity was in the order of 0.2 ppb. This value can be strongly enhanced by selecting another, more polar fibre coating.

In conclusion, the results obtained by us and by others indicate that SPME is a valid method for the analysis of target pesticides in water samples. Accuracy and reproducibility are good. Moreover, the data for the analysis of volatile pollutants in water samples (Nilsson *et al.*, 1996) were also excellent, which suggests that SPME could be a 'universal sampling method' for screening water pollution. It is important to note, however, that calibration is of utmost importance, which means that the method is restricted to the quantitative analysis of target compounds.

10.4.5 Static and dynamic sorptive extraction of phenols in water samples (Haghebaert *et al.*, 1996; Baltussen *et al.*, 1998e)

10.4.5.1 Introduction
Phenols occur widely in our environment (Tesarova and Pacarova, 1983). Because of their toxicity and persistency, monitoring of this class of pollutants in aqueous samples is required at or below the µg/l level. Adequate enrichment from the water sample is needed, followed by chromatographic analysis with sensitive and selective detection. Many enrichment techniques are available for the preconcentration of these analytes before the analytical quantification by gas or liquid chromatography. Classical LLE has been used, which is a simple and straightforward technique (Kopecni *et al.*, 1989). However, because of the well-known drawbacks, such as large consumption of organic solvents, LLE has increasingly been replaced by solid phase extraction (SPE). For the latter technique, a variety of adsorbent phases is available, including C8 and C18 modified silicas (Bao *et al.*, 1996), styrene-divinylbenzene co-polymers (Pocurrull *et al.*, 1994), graphitized carbon blacks (Rudzinski *et al.*, 1995) and XAD resins (Crespin *et al.*, 1997). Though preconcen-

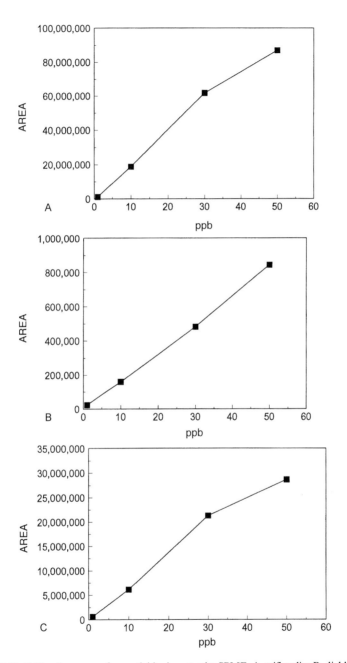

Figure 10.10 Calibration curves for pesticides in water by SPME. A, trifluralin; B, dichlorvos; C, aldrin.

270 EXTRACTION METHODS IN ORGANIC ANALYSIS

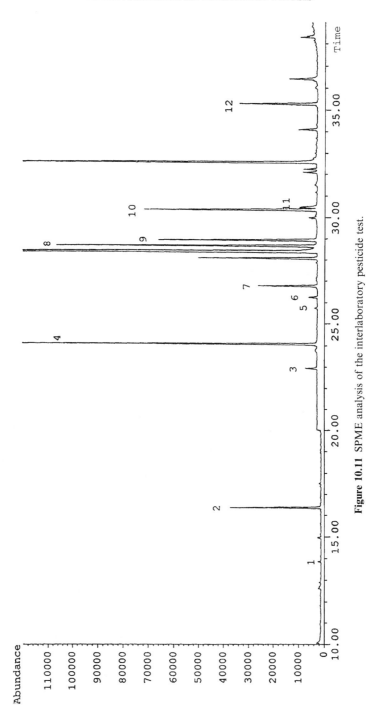

Figure 10.11 SPME analysis of the interlaboratory pesticide test.

tration of the phenols is possible, it does not generally provide high and consistent recoveries due to the high polarity of this class of solutes. This is especially true for phenol itself, which cannot be extracted quantitatively by any of the above-mentioned methods. Therefore, many authors have relied on the conversion of the phenols into less polar solutes. Phenols are easily acylated with acetic acid anhydride (Soniassy et al., 1994), facilitating both extraction and GC-MS analysis at low concentration levels.

The solventless techniques, SPME and automated dynamic sorptive extraction/thermal desorption in combination with CGC/MS will be described for the determination of phenols in aqueous samples.

10.4.5.2 Static sorptive extraction
Introduction. A water sample was received from the environmental organization of the Flemish Government (VITO, Belgium), with the request to determine the phenolic compounds in the sample in order to finalize our accreditation for water analysis. The analysis was performed by the classical and time-consuming method of LLE with DCM (Soniassy et al., 1994) and by SPME. SPME has already been evaluated for the analysis of phenols by Buchholz and Pawliszyn (1993, 1994). The data obtained by both methods are presented and a foolproof and accurate SPME procedure is described.

Experimental. SPME was performed manually using an SPME holder and fibres of 100 μm PDMS or 85 μm PA from Supelco Inc. (Bellefonte, PA, USA). Experiments were performed on a blind sample prepared by the Flemish reference laboratory VITO (Belgium). Three bottles of 1 l water adjusted to pH 3 and contaminated with unknown phenols were received. The first bottle was used to identify the phenols. The phenols were extracted by a classical procedure consisting of LLE with DCM (3 × 70 ml), after conversion of the phenols into their acetates and capillary GC-MS analysis (Soniassy et al., 1994). The internal standards, 2-methylphenol and 2,4,6-trichlorophenol, phenolic solutes which were not present in the blind sample, were added to the second bottle in concentrations of 19.2 and 21.0 μg/l, respectively. The classical procedure was then repeated to yield accurate quantitative data for the phenols identified. The response factors were determined by spiking a blank water sample with known concentrations of identified and standard phenols. After addition of the internal standards, 10 ml samples were transferred from the third bottle into 20 ml headspace vials for SPME. The samples were analyzed before and after *in situ* derivatization. Derivatization was carried out by adding 0.5 g potassium carbonate and 0.5 ml acetic acid anhydride to the 10 ml sample, followed by stirring for 15 min. Before sampling, the fibre was conditioned by placing it in an inlet of a GC instrument at 250°C for 3 min, at a flow rate of 50 ml/min helium. For analysis of both underivatized and derivatized samples, the fibre was introduced into the liquid and extraction was performed during 45 min under continuous stirring with a Teflon-coated magnetic stirrer at roughly 1000 rpm. After sampling, the fibres were desorbed in the split/splitless inlet of the GC instrument. The instrumental configuration and analytical conditions were as follows:

Instrumental configuration:

Gas chromatograph	HP 5890
Inlet	Split/splitless
Detector	MSD HP 5972
Liner	0.75 mm ID SPME insert (Supelco)
Data handling	Chemstation (DOS Series)
Column	30 m × 0.25 mm ID × 0.25 μm SPB-5 (Supelco)

Experimental conditions:

Inlet temperature	250°C
Thermal desorption	5 min
Injection mode	Splitless
Splitless time	5 min
Purge flow	50 ml/min
Carrier gas	Helium
Head pressure	60 kPa
Carrier gas mode	Constant pressure
Oven temperature	50°C (1 min)–10°C/min–280°C.
Detector	Operated in the ion-monitoring mode

Results and discussion. The phenolic compounds identified in the blind sample were 3-methylphenol (peak 1), 2,3,5-trimethylphenol (peak 2), 3,4-dichlorophenol (peak 3), 2,3,5,6-tetrachlorophenol (peak 4) and pentachlorophenol (peak 5). After addition of the internal standards, the areas were compared to those obtained by analysing a blank water sample spiked with known concentrations of both identified and standard phenols. The quantitative data of the LLE procedure together with the ions selected for quantitation by ion-monitoring MS are listed in Table 10.5.

In order to compare the performance of SPME on the PDMS and PA fibres for the underivatized and the *in situ* acetylated phenols, the critical steps of the SPME process, namely absorption time, desorption temperature and time, and stirring conditions, were kept constant. For the derivatized phenols on the PDMS fibre, the selected conditions, i.e. sampling for 45 min at 1000 rpm and desorption at 250°C during 5 min,

Table 10.5 Quantitative data by LLE and SPME

Peak	Compound	Ions	LLE	SPME	RSD%	% on fibre*
IS1	2-Methylphenol	108,107				0.3
1	3-Methylphenol	108,107	23.2	24.4	7.4	0.4
2	2,3,5-Trimethylphenol	121,136	18.8	19.5	7.2	0.2
3	3,4-Dichlorophenol	162,164	30.7	26.0	13.0	2.8
IS2	2,4,6,-Trichlorophenol	196,198				9.8
4	2,3,5,6-Tetrachlorophenol	232,268	18.9	18.6	8.3	25.0
5	Pentachlorophenol	266,268	10.5	11.2	6.9	31.6

*: for acylated phenols on PDMS. Abbreviations: LLE, liquid-liquid extraction; SPME, solid phase microextraction; RSD, relative standard deviation; PDMS, polydimethyl siloxane.

were found to provide the best results, also taking time into consideration. Under those conditions, the carry-over was less than 1% for both fibres, as ascertained by making a second desorption of the fibres.

Figure 10.12 shows the analyses, recorded to the same scale, of the sample on the PDMS fibre without (A) and after derivatization (B), and on the PA fibre without (C) and after derivatization (D). The identities of the peaks were elucidated from the data obtained by the classical sample preparation procedure.

The following conclusions can be drawn,

(A) SPME on PDMS—underivatized sample: None of the phenols was detected because of their strong hydrophilic nature.
(B) SPME on PDMS—derivatized sample: The extraction yields drastically increased and the enrichment was a function of the hydrophobicity. The high molecular weight solutes were extracted with high yield, e.g. IS2, peaks 4 and 5.
(C) SPME on PA—underivatized sample: The recoveries were much higher compared to the PDMS fibre for the underivatized phenols, which can be explained by the more polar nature of the fibre. This was confirmed by the decrease in function of the molecular weight or hydrophobicity. Pentachlorophenol was almost absent. Most of the solutes exhibited tailing, rendering quantification less accurate.
(D) SPME on PA—derivatized sample: The recoveries for the high molecular weight substances increased compared to SPME on PA for underivatized solutes, which can be explained by a decrease in polarity because of derivatization. Compared to the derivatized sample on the PDMS fibre, however, extraction efficiencies were much lower for high molecular weight solutes.

SPME on PDMS fibres, in combination with acylation, provided the best results, not only in terms of recovery but also in terms of chromatographic analysis. The quantitative data obtained in this manner

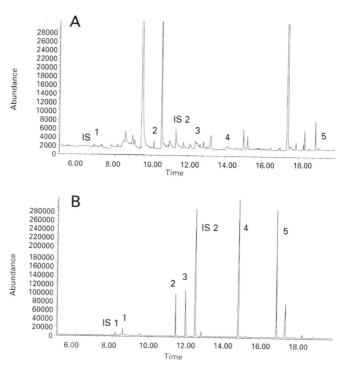

Figure 10.12 SPME of phenols in water. A, PDMS fibre for underivatized sample; B, PDMS fibre for derivatized sample; C, PA fibre for underivatized sample; and D, PA fibre for derivatized sample. Abbreviations: PDMS, polydimethylsiloxane; PA, polyacrylate.

for the blind sample are listed in Table 10.5 together with the percentage relative standard deviation (RSD%) values (n = 5). The similarity between LLE and SPME is good, and the repeatability values are in agreement with other SPME determinations and acceptable at these trace levels. The exact figures on the phenol concentrations are unknown to the authors. Based on the data presented in Table 10.5, the accreditation was, however, granted. The sensitivity and linearity of SPME on PDMS after derivatization were measured for pentachlorophenol. The sensitivity is in the order of 10 ppt with a linearity of at least 5.10^3. Figure 10.13 shows the ion-monitoring analysis (m/z 266) of 0.1 ppb pentachlorophenol.

10.4.5.3 Dynamic sorptive extraction
Introduction. As an alternative to SPME with a PDMS fibre, *in situ* derivatized phenols were also enriched from water onto a packed PDMS trap, followed by thermal desorption. The advantages over SPME

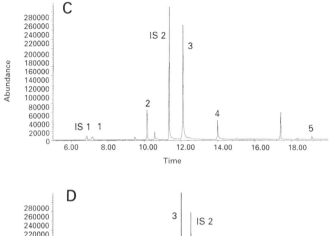

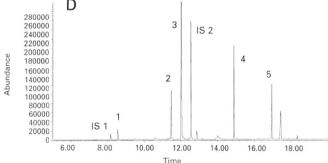

Figure 10.12 (Continued).

include total sorption, which results in high sensitivity and easier quantification.

Experimental. The phenols were obtained from Supelco (Bellefonte, PA, USA) and a standard solution was prepared in ethyl acetate. Spiked water samples were prepared in HPLC grade water by addition of the appropriate amount of the ethyl acetate solution. The following instrumentation and conditions were used:

Instrumental configuration:
Gas chromatograph   HP 6890
Inlet   Gerstel Thermodesorption System (TDS-2)
                         Gerstel TDS-L automatic liquid sampler
Detector   MSD HP 5972
Data handling   Chemstation (Windows Series)
Column   30 m × 0.25 mm ID × 0.25 µm HP-5MS

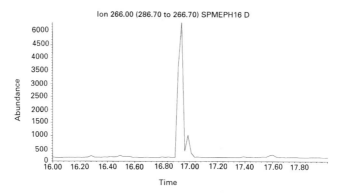

Figure 10.13 SPME for pentachlorophenol at the 0.1 ppb level.

Experimental conditions:

TD temperature	250°C
Thermal desorption	5 min
Injection mode	Splitless
Desorption flow	250 ml/min
Carrier gas	Helium
Head pressure	60 kPa
Carrier gas mode	Constant pressure
Oven temperature	40°C (2 min)–15°C/min–300°C
Detector	Operated in the ion-monitoring mode

The instrumentation used was the same as that described previously (Baltussen *et al.*, 1998f) consisting of an autosampler for automated sample pretreatment (Thermodesorption system automatic liquid sampler (TDS-L); Gerstel, Müllheim a/d Ruhr, Germany), a thermal desorption unit (TDS-2, Gerstel) and an HP 6890 GC coupled to a HP 5972 MSD (Hewlett Packard, Little Falls, DE, USA). An important feature of the system is a backflush adapter that is installed at the bottom of the cryotrap. Activating the backflush line allows backflushing of the cryotrap, thus preventing water vapour from entering the analytical column.

The procedure started with derivatization of the phenols in the water sample. One millilitre of acetic acid anhydride and 1 g of sodium hydrogen carbonate were added to a 20 ml water sample. After brief shaking, the mixture was allowed to stand for 15 min to complete the reaction. Subsequently, 10 ml of the sample was pumped through the PDMS cartridge. When sampling was complete, the cartridge was washed with 10 ml pure water to remove excess reagent, solubilized salts and other undesired matrix compounds. Prior to thermal desorption, the

Table 10.6 Ions selected for selected ion-monitoring mode (SIM) analysis

No.	Component	Ion 1	Ion 2	Group
1	Phenol	94	136	1
2	4-Methylphenol	108	150	2
3	2-Chlorophenol	128	170	3
4	2,6-Dimethylphenol	122	164	3
5	2-Ethylphenol	122	164	3
6	4-Isopropylphenol	136	178	4
7	2,4-Dichlorophenol	162	204	5
8	2,3,5-Trimethylphenol	136	178	5
9	2,4,6-Trichlorophenol	196	238	6
10	Pentachlorophenol	264	306	7

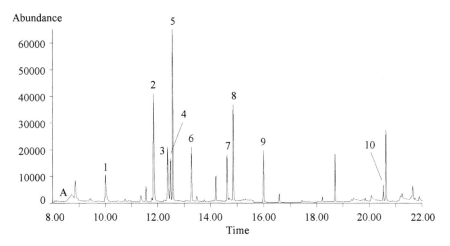

Figure 10.14 Dynamic sorptive extraction on PDMS of phenols in water at the 0.1 ppb level.

cartridge was dried at ambient temperature for 25 min and at 50°C for 5 min. Both steps were carried out under a nitrogen flow of 300 ml/min. Finally, the phenols were thermally desorbed at 225°C for 5 min, cryofocussed at −150°C and splitless injected onto the capillary GC column. The mass selective detector was operated either in the full scan mode by scanning from 40 to 300 amu, or in the selected ion-monitoring mode (SIM) by monitoring two ions per component as indicated in Table 10.6.

Results and discussion. A tap water sample was spiked at a level of 0.1 µg/l with the underivatized phenols. Twenty millilitres of the spiked tap water sample was derivatized as described previously and 10 ml was enriched onto a PDMS cartridge. Therefore, a total of 1 ng per compound was injected onto the column by the SE/TD technique. Figure 10.14 shows the result of the SIM analysis. Table 10.7 shows the recoveries of the test analytes calculated by comparison with a direct injection.

From Table 10.7, it is clear that most analytes are quantitatively enriched on the PDMS material. Only the first two analytes (phenol and 4-methylphenol) are partially lost. The sensitivity of the method is very good, as can be deduced from the abundances in the SIM trace for 0.1 μg/l. For a signal-to-noise ratio of 8, limit of detection (LOD) values are 1–10 ng/l!

Table 10.7 Recovery and RSD% (n = 3) for the analysis of tap water spiked at 0.1 ppb with the phenols

No.	Component	Recovery %	RSD %
1	Phenol	72	5
2	4-Methylphenol	64	16
3	2-Chlorophenol	95	11
4	2,6-Dimethylphenol	107	7
5	2-Ethylphenol	101	2
6	4-Isopropylphenol	109	15
7	2,4-Dichlorophenol	103	5
8	2,3,5-Trimethylphenol	97	3
9	2,4,6-Trichlorophenol	94	9
10	Pentachlorophenol	98	5

Abbreviations: RSD, relative standard deviation.

10.4.6 Solid phase extraction of triazines from water samples on Empore disk cartridges followed by large volume injection (Gerstel Aktuell, 1977)

10.4.6.1 Introduction

For environmental applications, SPE is at present usually carried out using cartridges containing 100 mg to 1 g nonpolar phase, such as octadecyl silica (ODS) or polystyrenedivinylbenzene (PSVB) co-polymer. These cartridges perform very well for nonpolar and medium polarity solutes, provided that the sampling rate is low to prevent breakthrough (typically 10–20 ml/min, which results in a sampling time of 1 h for a 1 l sample), and that drying of the cartridge is complete, which can take a long time (30 min). As an alternative to SPE catridges, extraction disks (Empore TM Disks, 3M, St. Paul, MN, USA) were introduced allowing higher sampling flow rates (1 l in 10 min) and reduced drying times. These disks (47 mm diameter) are made by impregnating small particle size ODS material (12 µm compared to 40 µm in conventional SPE cartridges) in a polytetrafluoroethane (PTFE) fibril network. For desorption, relatively high solvent volumes (typically 10 ml) are required and a concentration step is needed to reach the detection limits.

Recently, a new approach to SPE in environmental monitoring was presented, namely the use of Empore TM disk cartridges (Sandra *et al.*, 1996a). The advantages of disk cartridges (10 mm diameter) are the fast sampling rate, the short drying period, and the low solvent volume needed for desorption. These features make them ideally suited for on-site and in-field sampling. For environmental samples, their applicability appears to be limited because the sample loadability is reduced compared to classical disks or cartridges (10–50 ml compared to 1 l). Moreover, detection limits set by official organizations seem difficult to achieve. The European Community (EC) directive for pesticides in water, for example, implements the determination of 1 µg/l concentrations in water samples and even of 0.1 µg/l levels in drinking water. With classical SPE methods, these detection limits can be reached by extraction of a 1 l sample, concentration to 1 ml (enrichment factor = 1000), and injection of a 1–2 µl sample into the GC instrument using splitless or cool on-column injection. Using disk cartridges, this enrichment factor is difficult to obtain. Recent developments in sample introduction into the CGC, namely large volume injection (LVI) via a programmable temperature vaporizing (PTV) injector, have helped to overcome this problem. By operating a PTV in the solvent venting mode, large volumes can be injected (Staniewski and Rijks, 1993). By SPE of a 25 ml sample on a disk cartridge, desorption with 500 µl solvent and direct injection (i.e. without concentration) of 40 µl extract, the same amount is entering the column as in classical SPE, thus giving the same sensitivity. This is illustrated in Table 10.8.

Table 10.8 Comparison of classical SPE and SPE using disk cartridges

	Classical SPE	SPE on disk cartridge
Sample concentration	0.1 ppb	0.1 ppb
Sample volume	1 l	25 ml
Extraction	1 g ODS cartridge or 47 mm ODS disk	10 mm ODS disk cartridge
Elution	10 ml	0.5 ml
Final volume	1 ml	0.5 ml
Final concentration	100 ng/ml	5 ng/ml
Injection volume	2 µl	40 µl
Injection amount	200 pg	200 pg

Abbreviations: SPE, solid phase extraction; ODS, octadecyl silica.

The sample preparation time is strongly reduced and the concentration step is avoided. SPE on disk cartridges in combination with large volume PTV injection and capillary GC–MS will be demonstrated by the analysis of the widely-used triazine pesticides in water.

10.4.6.2 Experimental
Solid phase extraction was performed on 10 mm/6 ml ODS Empore TM disk cartridges. The disk cartridges were conditioned with 0.5 ml methanol and 1 ml water. Twenty-five millilitres of water was passed through the cartridge at a flow rate of 50 ml/min. After drying under vacuum, desorption was performed directly in a 2 ml autosampler vial with 0.5 ml ethyl acetate. Water samples were spiked with 1 and 0.2 µg/l of the triazines listed in Table 10.9. The instrumental configuration and analytical conditions were as follows:

Instrumental configuration:

Gas chromatograph	HP 5890
Inlet	PTV–Gerstel (Müllheim a/d Ruhr, Germany)
Detector	HP 5971 MSD
Data handling	Chemstation (DOS series)
Column	30 m × 0.25 mm ID × 0.25 µm HP-5MS

Experimental conditions:

Inlet temperature	30°C–2°C/s–70°C–24 s–10°C/s–250°C–60 s
Injection volume	40 µl at 100 µl/min
Injection mode	Solvent venting mode
Purge valve	Initially ON, OFF at 0.5 min, ON at 1.5 min
Carrier gas	Helium
Head pressure	50 kPa
Carrier gas mode	Constant pressure
Flow, velocity	1 ml/min, 30 cm/s
Oven temperature	50°C (1 min)–10°C/min–300°C

10.4.6.3 Results and discussion
The analysis of a blank water sample spiked at the 1 µg/l level is presented in Figure 10.15A. All triazines are easily detected, with good peak shape. Because of solvent venting in the PTV inlet, peak distortion by band-broadening in space is absent and no retention gap is needed for 40 µl injection. The injection speed was optimised at 100 µl/min in order to avoid, on the one hand, liner overflow (liquid sampling flowing into the column if injection is too fast) and, on the other, loss of solutes due to drying of the liner (injection too slow). The optimum injection speed can

Table 10.9 Peak elucidation, retention times, ions and quantitative results

Peak no.	Triazine	t_R min	Quant. ion	Qual. ion	Recovery %	RSD %
1	Desisopropylatrazine	15.39	173	158	39	13.8
2	Desethylatrazine	15.55	187	172	87	12.2
3	Simazine	16.45	201	186	107	9.8
4	Atrazine	16.57	200	215	94	8.2
5	Propazine	16.66	214	229	79	7.4
6	Terbutylazine	16.89	214	229	84	3.0
7	Sebutylazine	17.54	200	229	101	9.9
8	Metribuzine	18.02	198	214	22	2.5
9	Prometryn	18.40	241	184	74	64
10	Terbutryn	18.69	226	241	68	50
11	Cyanazine	19.18	225	240	91	15.4

Recovery measured at the 1 ppb level. RSD measured at the 1 ppb level (n = 5). Abbreviations: RSD, relative standard deviation; t_R, retention time; Quant., quantification; Qual., qualification.

be calculated using the LVI calculator software. The retention time, the quantitation ion, the qualifier ion, the recovery and the RSD on peak areas for five experiments for the different triazines are presented in Table 10.9. The recovery was measured by comparing the peak areas for the analysis of the sample extracts (1 µ/l level) with a direct analysis of a diluted standard solution. For most compounds, recoveries were 70–110%. Recoveries for desisopropylatrazine and metribuzin were lower, but this is known on ODS. PSDVB performs much better in this respect (Sandra *et al.*, 1995). For most solutes the RSDs were > 10%, which is excellent for these polar pesticides. This value, of course, includes both the SPE method and the deviation of the large volume injection. Experiments with other classes of compounds, such as PAHs, phthalates, and organochloro-pesticides, showed higher recoveries and better RSDs. The sensitivity of the method is illustrated in Figure 10.15B, showing the analysis of an extract from a water sample spiked at the 0.2 ppb level. The extracted ion chromatograms for the first four compounds show that the detection limits set by the EC can easily be reached with ion extraction, while specific ion-monitoring allows determinations down to the ng/l level.

10.4.7 Solid phase extraction of pesticides from water samples followed by supercritical fluid desorption and supercritical fluid chromatography (Sandra *et al.*, 1995a)

Sample preparation is the first priority for application of reliable and cost-effective automation. SPE is increasingly replacing LLE for the enrichment of organic micropollutants from water samples, e.g. PAHs, organochloro-pesticides, phenols, phenylureas, etc. In addition to

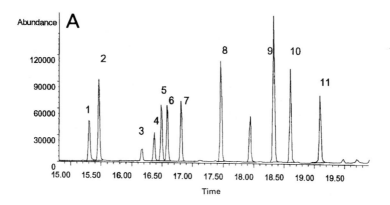
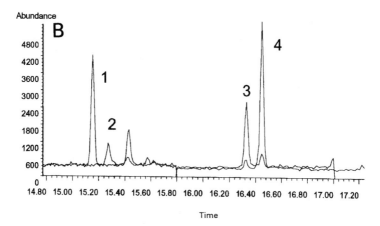

Figure 10.15 SPE followed by large volume injection of triazines in water. A, At the 1 ppb level; B, At the 0.2 ppb level.

providing a better performance compared to traditional methods, SPE has the distinct advantages of easy automation and the possibility of on-line coupling with the separation system. The literature is well-documented on hyphenation of SPE with different chromatographic techniques (Sandra et al., 1995b; Vreuls et al., 1994; Noij and van der Kooi, 1995; Sandra et al., 1996b), and a full range of adsorbents have been applied, the most important of which are ODS and PSDVB resins.

Until recently, SFC was not regarded as either a quantitative or a trace analysis technique. However, the introduction of reliable instrumentation and the revival of packed column SFC has changed this situation and opened the technique for precolumn hyphenation with the sample preparation step.

The combination of solid phase extraction (SPE) and carbon dioxide desorption (SFE)-supercritical fluid chromatography (SFC) has some advantages over its CGC (SPE-CGC) and HPLC (SPE-HPLC) counterparts. Compared to CGC, there is no need for complex interfacing to volatilize and remove the desorption fluid, while, compared to HPLC, the selectivity of desorption can be better controlled.

Two SPE-SFC-DAD combinations have recently been described (Medvedovici *et al.*, 1997). The first system, based on the Gilson 233 XL sample preparation station (Figure 10.16), is a simple single cartridge approach which can be applied if the sample load is rather low, whereas the second system, based on the Merck-Hitachi automatic SPE preparator (OSP-2A), is a multi-cartridge system which makes it possible to monitor unattended organic micropollutants in 16 water samples (Figure 10.17).

The best adsorbents for SPE-SFE-SFC, in terms of recoveries (adsorption), peak shapes, carry-over (desorption kinetics) and total chromatographic profile (selectivity of desorption), are PSDVB resins.

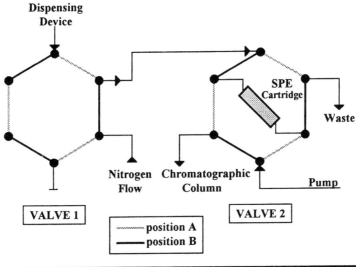

Operation	Position Valve 1	Position Valve 2
Conditionning, Loading, Washing	A	B
Drying	B	B
Injection	A	A

Figure 10.16 Single cartridge SPE-SFC-DAD system. Abbreviations: SPE, solid phase extraction; SFC, supercritical fluid chromatography; DAD, diode array detector.

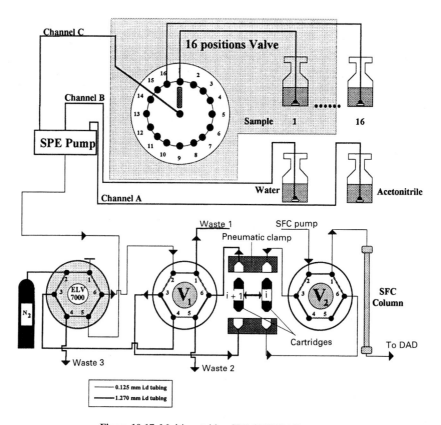

Figure 10.17 Multi-cartridge SPE-SFC-DAD system.

LiChrolut EN with 60 μm particle size (Merck) is a highly hydrophobic material with excellent adsorption characteristics for polar herbicides and pesticides.

Besides the nature of the adsorbent, the parameters controlling the adsorption are the sample loading flow rate and the sample volume. This was studied in depth by Medvedovici and co-workers (1997). The most critical step in the SPE-SFC procedure is, however, the drying step. In fact, without drying the adsorbent between adsorption and desorption, the SPE-SFC hyphenation is not working at all because of excessive peak distortion. Residual water traces result in a desorption process with two nonmiscible fluids (CO_2 and water) which gives rise to peak-broadening and splitting. For 60 μm particles, a drying period of 15 min is optimal. Table 10.10 presents recoveries at the 20 ppb level for 30 pesticides together with the %RSDs for a series of eight experiments. For comparison, the %RSDs are also presented for five direct injections of the same absolute amounts via a 5 µl sample loop. With the exception of chloroxuron and chlorizadone, the recoveries are very good, while for all pesticides, the %RSDs fall in the range of the chromatographic %RSDS, illustrating the excellent repeatability of SPE-SFC-DAD.

An interesting feature of CO_2 desorption is its selectivity. Only the solutes which are soluble in the fluid will be desorbed and enter the column. This means that the polar and ionic solutes which often occur in real world samples, e.g. humic acids, fulvic acids, lignine, detergents, etc.,

Table 10.10 Recoveries at the 20 ppb level for pesticides by SPE-SFC-DAD

No.	Pesticide	RSD% for peak areas SPE-SFC-DAD	RSD% for peak areas SFC-DAD	Recovery %
1	Propham	1.5	1.6	94.0
2	Chlorpropham	2.5	1.3	93.7
3	Metolachlor	3.9	2.4	91.6
4	Propazine	0.5	2.3	100.8
5	Sebutylazine	1.3	1.4	90.6
6	Terbutylazine	2.2	0.6	103.2
7	Prometryn	0.9	2.7	103.8
8	Methazachlor	2.0	3.2	95.8
9	Atrazine	0.8	0.7	112.2
10	Metobromuron	2.2	1.5	88.5
11	Monolinuron	0.7	0.9	105.9
12	Terbutryn	2.3	1.4	98.1
13	Metribuzin	1.1	0.6	102.1
14	Simazine	0.5	0.5	106.1
15	Linuron	1.6	0.4	102.6
16	Cyanazine	1.1	0.4	105.4
17	Methabenthiazuron	0.6	0.7	102.3
18	Desethylatrazine	0.3	0.6	107.5
19	Bromacil	1.4	3.1	104.1
20	Crimidine	0.4	0.8	108.5
21	Desisopropylatrazine	0.4	1.0	105.7
22	Isoproturon	0.2	0.9	106.0
23	Fenuron	0.4	1.3	99.3
24	Chlortoluron	0.8	0.8	104.3
25	Diuron	0.9	0.6	101.6
26	Matamitron	1.2	0.8	90.6
27	Hexazinone	1.0	1.0	90.9
28	Chloroxuron	4.3	1.9	45.3
29	Methoxuron	0.9	0.9	98.8
30	Chlorizadone	4.2	0.9	51.2

Abbreviations: SPE, solid phase extraction; SFC, supercritical fluid chromatography; DAD, diode array detector; RSD, relative standard deviation.

in surface, ground and river waters, will not be introduced into the column. Compared to SPE-LC-DAD, where the desorption fluid is highly hydrophilic and desorbs some of these solutes as well, clean chromatograms with flat baselines are obtained. This renders qualitative and quantitative analysis easier. This is illustrated in Figure 10.18, showing the analysis of water collected from the river Lys in July 1995. In this particular case, in order to emphasize the selectivity of CO_2 desorption and to allow comparison with published SPE-LC-DAD chromatograms, a 10 μm ODS cartridge was used. Because of their retention times and DAD spectra (inserts of Figure 10.18), atrazine and diuron could easily be elucidated and quantified. The concentrations were 1.38 ppb for atrazine and 0.95 ppb for diuron, with RSD% of 4.8 and 6.1, respectively, for triplicate analysis. Desethylatrazine was present at 0.28 ppb.

The analytical conditions were as follows:

Instrumental configuration:
SPE-SFC-DAD	OSP 2A (Merck)–HP SFC
Data handling	Chemstation (DOS series)
Column	20 cm × 4.6 mm ID Hypersil 5 μm (HP)

Experimental conditions:
Flow rate	2 ml/min
Temperature	40°C
Pressure programme	100 bar–5 bar/min–250 bar
Modifier	Methanol 1%–0.5%/min–21%

10.4.8 Gas phase extraction of PAHs and PCBs from soil samples (Wormann and Hoffmann, 1994)

10.4.8.1 Introduction

Thermodesorption is a well-known technique for the analysis of volatile contaminants in air after enrichment by adsorption on porous polymers or sorption on PDMS. The same operation principle can also be adopted for the direct thermal desorption of volatiles within a wide boiling range directly from soil samples. The combination of thermal desorption with intermediate cryofocusing in the insert liner of a PTV injector is a reliable and fast method for the determination of solutes, even with boiling points as high as, for example, PAHs and PCBs. The principle will be illustrated with the analysis of PAHs and PCBs in a contaminated soil.

10.4.8.2 Experimental

No other sample preparation than crushing the soil with a jaw crusher to a size of approximately 4 mm was needed. Five hundred milligrams of the

ENVIRONMENTAL APPLICATIONS

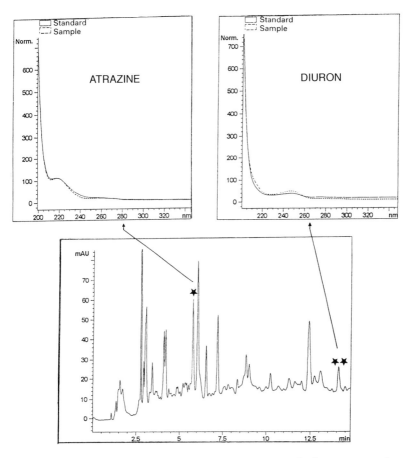

Figure 10.18 SPE-SFC-DAD of river Lys water. ———, standard; - - - - - , sample.

soil was then introduced into a thermal desorption tube and a solution of 9,10-diphenylanthracene and PCB 209 standard was added for quantitation of the PAHs and the PCBs, respectively. The tube was then placed into the thermal desorption unit. The other conditions were:

Instrumental configuration:
Gas chromatograph   HP 5890 series II
Thermodesorption system   TDS 2 (Gerstel)
PTV injector   CIS 3 (Gerstel)
Detector   HP 5972

Experimental conditions:
TDS 2:

Pneumatic mode	Split
Temperature	−50°C (1 min)–60°C/min–350°C (20 min)
Transfer line	350°C

CIS 3:

Liner	Deactivated glass wool
Carrier gas	Helium
Pneumatic mode	Solvent venting
Vent pressure	30 psi, 150 ml/min
Vent time	0.01 min
Splitless time	2 min
Temperature	−150°C–12°C/s–400°C (10 min)

GC:

Column	50 m × 0.25 mm × 0.25 µm BB-5 (J&W)
Carrier gas	Constant pressure
Temperature	60°C (1 min)–10°C/min–250°C–5°C/min–320°C (20 min)

MSD:

MSD–PAHs–Scan mode	Ion extraction
MSD–PCBs–Scan mode	Ion-monitoring (6 masses per group)

10.4.8.3 Results and discussion

This method provides all qualitative and quantitative information about the PAHs and PCBs down to the ppb level. Figure 10.19 shows the total ion chromatogram (A), the extracted ion chromatogram for the PAHs (B), and the ion-monitoring chromatogram (C) for the PCBs. The latter was obtained by a second analysis of the same sample. The peaks marked with an asterisk are the internal standards. The only slight disadvantage of the method concerns the small sample amount taken, which is often considered as being non representative from a pollution site. The fact that the total extraction time is only 30 min and that the system is fully automated makes it possible, however, to run a number of samples to verify the homogeneity.

10.4.9 Static sorptive extraction of organometallics from environmental samples

10.4.9.1 Introduction

The toxicity of metal-containing organic compounds is highly dependent on their chemical structure. Subtle differences in analogue molecules at

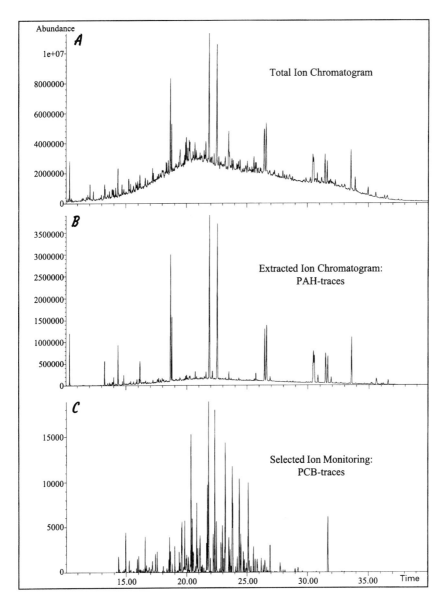

Figure 10.19 Direct thermal desorption of contaminated soil. A, Total ion chromatogram; B, PAH extracted ion chromatogram; C, PCB selected ion chromatogram. Abbreviations: PAH, polycyclic aromatic hydrocarbons; PCB, polychlorinated biphenyl.

first glance can have severe consequences on their chemical toxicity (Craig, 1986). As a consequence, speciation of organometallic compounds has gained increasing importance and several hyphenated techniques, i.e. a combination of a separation technique with an element specific detection system, have been described in the literature (Hill et al., 1993; Smits, 1994; Lobinski, 1997). Since the introduction of commercial capillary gas chromatography-microwave induced plasma-atomic emission detection systems (CGC-MIP-AED), organometallic analysis can be performed routinely. In addition, the coupling of CGC to inductively-coupled plasma mass spectrometry (ICPMS) has proved to be a highly sensitive and selective method for organometallic speciation (De Smaele et al., 1995, 1996a, 1996b). Both GC and ICPMS can work independently and can easily be coupled within a few minutes by means of a transfer line. Therefore, the hyphenation of these instruments is even more attactive than the CGC-MIP-AED system for environmental laboratories, where ICPMS is often the method of choice for routine determination of total metal concentration.

Since most organometallic components of tin, mercury and lead occur in nature in rather involatile ionic species, these compounds have to be derivatised into volatile solutes prior to GC analysis. The sample preparation for organometal speciation by CGC has been drastically simplified by the introduction of an aqueous *in situ* derivatization by Ashby and Craig (1991). Derivatization with sodium tetraethylborate, $NaBEt_4$, has been investigated for a wide range of organometallic compounds, such as organolead, -mercury, -cadmium, -tin and -selenium. Recently, De Smaele and co-workers (1998) reported on the possibilities of *in situ* propylation as a novel aqueous derivatisation method, with which even the ethyl derivatives of Hg and Pb can be volatilized. The advantages of this derivatisation technique are that alkylation takes place in the aqueous phase and extraction can be performed simultaneously. In addition, it is compatible with modern extraction methods, such as purge and trap (Ceulemans and Adams, 1996), or sample preparation techniques, such as microwave-assisted extraction (Sprunar et al., 1996). SPME in combination with CGC-ICPMS was used for the first time by Moens and co-workers (1996) for the simultaneous extraction and determination of organotin, -mercury and -lead compounds. The application of SPME–CGC–ICPMS will be illustrated with the analysis of some environmental samples.

10.4.9.2 Experimental
The instrumental configuration and analytical conditions were as follows:

ICPMS	Perkin Elmer Sciex Elan 5000
RF power	1250 W
Sampling depth	10 mm
Carrier gas flow rate	1.10–1.25 l/min
Auxiliary gas flow rate	1.20 l/min
Plasma gas flow rate	15 l/min
Sampling cone/aperture diameter	Ni/1 mm
Skimmer cone/aperture diameter	Ni/0.75 mm
Dwell time	30–50 ms (depending on number of nuclides to be measured) 10 ms (^{126}Xe)
Transfer line	Home-made (De Smaele et al., 1996) 250°C
Gas chromatograph	Perkin Elmer Autosystem
Column	FSOT, methylsilicone 30 m; 0.25 ID; $d_f = 0.25\,\mu$m
Injection technique	Splitless
Injection temperature	250°C
Temperature programme	60°C (1 min)–20°C/min–200°C (0.5 min)
Carrier gas/inlet pressure	Xe/H$_2$ (1/99 mixture); 30 psi

ICPMS is well known for its multi-element capabilities, and coupled to a chromatographic system this feature can be fully exploited. In the case of organometallic speciation with CGC-ICPMS, the most abundant isotopes of the different elements, leading to the most sensitive signals, e.g. ^{120}Sn, ^{202}Hg and ^{208}Pb for Sn, Hg and Pb, respectively, can be chosen. In addition, ^{126}Xe, originating from Xe doped at a concentration of 1% (v/v) to the carrier gas, can be monitored continuously during the analyses and acts as an internal standard to correct for instrument instabilities, malfunctions or signal drifts during the GC analyses. A typical multi-element chromatogram is presented in Figure 10.20. Table 10.11 presents the abbreviations used.

In comparison with CGC-MIP-AED, detection limits with ICPMS are at least a factor of 10 superior, i.e. ICPMS instrumental detection limits for organo-Sn, -Hg and -Pb are in the order of 10–100 fg absolute as metal.

A SPME fibre holder for manual injections was used, with a 100 μm PDMS-coated fibre. Fifty millilitre glass vials closed with PTFE-coated rubber septa were used for sampling. Proper mixing of the sample

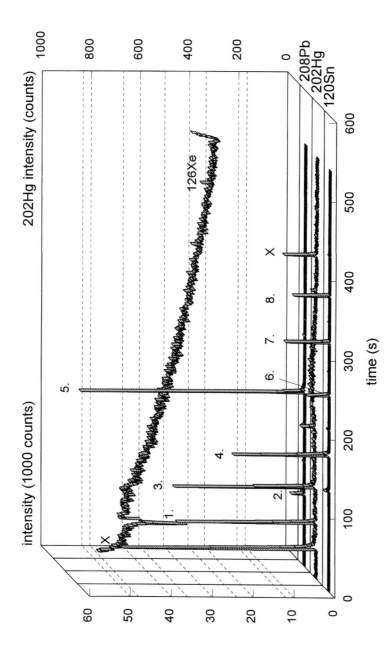

Figure 10.20 Chromatogram of an organometal standard extracted with SPME. 1, methylmercury (MM); 2, trimethyllead (TML); 3, inorganic mercury (diethylmercury); 4, inorganic tin (tetraethyltin); 5, inorganic lead (tetraethyllead); 6, monobutyltin (MBT); 7, dibutyltin (DBT); 8, tributyltin (TBT); and X, unknowns.

Table 10.11 Abbreviations for organometallics

Abbreviation	Organometallic
DBT	Dibutyltin
TBT	Tributyltin
TPT	Tripropyltin
MBT	Monobutyltin
MM	Methylmercury
TML	Trimethyllead

solutions during the SPME extractions was achieved with a magnetic stirrer.

10.4.9.3 Results and discussion

All environmentally relevant organometallic species of Sn, Hg and Pb occur in ionic form in nature. Therefore, they need to be derivatized into nonpolar volatile species by use of $NaBEt_4$ or $NaBPr_4$. Commonly, a 25 ml sample of a standard solution, buffered at pH 5.3 with NaOAc/HOAc is introduced to the vial by means of a syringe and 100–1000 µl of a 1% aqueous $NaBEt_4$ or $NaBPr_4$ solution is added. The vial is sealed and, immediately afterwards, the SPME fibre is exposed to either the sample headspace or the aqueous phase, depending on the sampling method. From classical LLE and CGC-ICPMS detection, it was found that the derivatization was completed in less than 10 min, which corresponds to the residence time of the fibre in the vial. Figure 10.21 shows the relative signals for different organotin, -mercury and -lead species measured with CGC-ICPMS after SPME (direct and headspace, 10 min extraction time) and splitless injection of an iso-octane extract. The signal intensities observed were normalized so as to correct for the different concentrations of solutions used for LLE (100 µg/l as metal) and SPME (2 µg/l). The sensitivity of headspace SPME is higher by a factor of up to 10 (MBT) when compared to direct SPME, and by a factor of up to 324 (MBT) when compared to LLE.

The derivatized organometallic species are completely apolar and poorly soluble in water and tend, therefore, to migrate to the headspace and then to the nonpolar PDMS fibre. This is highly advantageous to 'real life' sample analysis. Environmental samples are known to be very dirty, and low volatile organics, co-extracted in classical LLE, interfere in the separation of organometals.

The most abundantly used organometallic compounds are organotin, -mercury and -lead (Craig, 1986). Surface water samples were collected from industrial regions in the north east of Belgium (Limburg) and monitored for organotin, -mercury and -lead. The samples were acidified to pH 2 with concentrated HCl and stored in the dark in glass bottles at

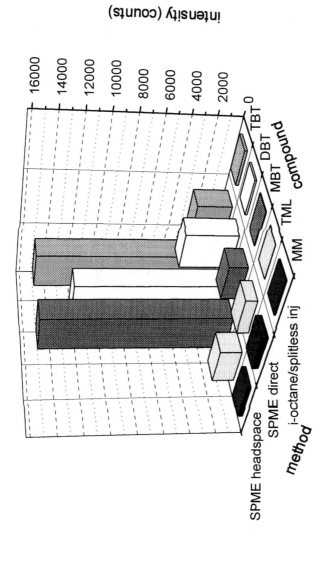

Figure 10.21 Relative sensitivity for methylmercury (MM), trimethyllead (TML), monobutyltin (MBT), dibutyltin (DBT) and tributyltin (TBT) for SPME headspace, SPME direct and liquid-liquid extraction.

4°C until analysis. The sample preparation was very simple and short. Twenty-five millilitres of water sample was pipetted into a sample vial and buffered with NaOAc/HOAc buffer (pH 5.3). NaBEt$_4$ was added, followed by SPME extraction for 10 min. The GC analysis takes about 10–12 min (including oven temperature re-equilibration) so that the sample preparation and analysis run almost concurrently. Figure 10.22 presents the concentrations of organometallic species found in water samples from different sites.

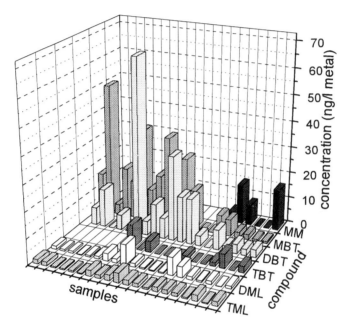

Figure 10.22 Concentrations of methylmercury (MM), trimethyllead (TML), dimethyllead (DML), monobutyltin (MBT), dibutyltin (DBT) and tributyltin (TBT) in water samples from different sites.

The concentrations are very low for all organometallic species and vary between 0.1 and 67.1 ng/l as metal. The average concentration of dibutyltin (DBT) is most abundant. The amount of tributyltin (TBT) is low (1.4–9.0 ng/l), since TBT tends to sorb onto the sediment and solid particles. These figures illustrate the very high sensitivity of the SPME technique. Twenty-five millilitres of water sample is sufficient, whereas for classical LLEs, at least 500 ml of water should be extracted and the solvent should be evaporated to a final volume of 500 µl.

The reliability of SPME-CGC-ICPMS for the determination of organotin and -mercury compounds was checked by the analysis of the standard reference material PACS-1 (marine sediment) and DORM-2 (dogfish muscle), both from the National Research Council, Canada

(NRCC). Approximately 0.2 g of sediment was weighed in a 50 ml glass vial and the internal standard, TPTOAc, was added. The organotin compounds were leached from the sediment matrix by adding 5 ml of HOAc and 5 ml of MeOH. The vial was closed and placed in an ultrasonic bath for 30 min. Subsequently, 250 µl of the supernatant was pipetted into another vial. Twenty-five millilitres of NaOAc/HOAc buffer at pH 5.3 was added. The vial was sealed and 1000 µl of 1% NaBEt$_4$ was added with a syringe. Subsequently, the SPME device was inserted and the PDMS fibre was exposed to the sample headspace for 10 min, followed by CGC-ICPMS analysis.

The fish tissue, on the other hand, had to be hydrolysed. The organomercury compounds are incorporated in the biological tissue matrix so that leaching alone cannot liberate the organometal species. Ten millilitres of 10% KOH was added to approximately 0.1 g of fish sample. Hydrolysis took place for 1–2 h under ultrasonic treatment. Fifty microlitres of the dissolved tissue was then pipetted into 25 ml of NaOAc/HOAc buffer (pH 5.3) followed by derivatisation, SPME extraction and CGC-ICPMS analysis. The results obtained with SPME-CGC-ICPMS are summarized in Table 10.12.

Table 10.12 Accuracy of SPME-CGC-ICPMS for certified reference materials

Compound	Headspace SPME ng/g metal	Certified values ng/g metal
NRCC PACS-1 Marine sediment		
MBT	750 ± 210*	280 ± 170
DBT	1060 ± 150	1160 ± 180
TBT	1220 ± 190	1270 ± 220
NRCC DORM-2 Dogfish muscle tissue		
MM	4280 ± 910	4470 ± 320

*: 95% confidence limit (n = 3). Abbreviations: MBT, monobutyltin; DBT, dibutyltin; TBT, tributyltin; MM, methylmercury; SPME, solid phase microextraction; CGC, capillary gas chromatography; ICPMS, inductively-coupled plasma mass spectrometry; NRCC, National Research Council of Canada.

As can be seen, all concentrations are in good agreement with the certified values except for monobutyltin (MBT). The certified value of MBT, however, is known to be too low (Moens et al., 1996). Recoveries of 95% were found for methylmercury (MM), and 91% and 97% for DBT and TBT, respectively. These figures prove that SPME is a reliable extraction technique for quantitative analysis.

10.4.10 Supercritical fluid extraction of PCBs from oils and fats

10.4.10.1 Introduction

For SFE extraction of PCBs from sediment, soil and sewage sludge, good results are obtained using pure CO_2 at a density of 0.75 g/ml, a temperature of 60°C and a flow rate of 2 ml/min for 30 min. Under these conditions, the PCBs are perfectly soluble in CO_2 and matrix effects are suppressed. Recoveries are similar to those obtained by classical methods of sample preparation. When PCBs have to be fractionated from fat matrices the extraction is much more difficult. The problem here is that extraction conditions developed for soil, sediment and sewage sludge samples yield high recoveries of lipids, and an additional clean-up on silica is required before chromatographic analysis. Otherwise, the co-extracted fat contaminates the inlet and/or the capillary column. Fractionation by size exclusion chromatography is normally applied to remove the PCBs from fat matrices. By fine-tuning the SF extraction conditions, it is also possible to obtain extracts with very low fat content that can be analyzed directly, i.e. without additional clean-up. Preliminary data have been published for the extraction of PCBs from seagull eggs and from milk powder (David *et al.*, 1992). The influence of the different parameters on the extraction of fat matrices has been studied in depth for the extraction of a certified cod liver oil sample (BCR, Brussels, Belgium).

10.4.10.2 Experimental

The oil was deposited on filter paper and extracted using different conditions.

The GC analyses were performed on a 50 m × 0.25 mm ID × 0.25 µm SE-54 column. Hydrogen was the carrier gas at 40 cm/s linear gas velocity, and the oven temperature was programmed from 80 to 150°C at 25°C/min, and then to 300°C at 5°C/min. The samples were injected in the splitless mode. The concentration of the individual congeners was calculated by the external standard method. The extraction recovery for each congener was calculated relative to the certified values. These values are 68 ppb (µg/kg) for PCB 28, 149 ppb for PCB 52, 370 ppb for PCB 101, 456 ppb for PCB 118, 938 ppb for PCB 153 and 282 ppb for PCB 180. The recoveries obtained for each congener were then averaged to obtain a single value giving the mean recovery obtained by SFE. The sample extracts were analyzed on the presence of co-extracted fat by hydrolysis and methylation with methanol/sulphuric acid (95:5) into the fatty acid methyl esters (FAMES) and analysis by CGC-flame ionization detection (FID). The fat concentration in the extracts, compared to the value obtained by direct hydrolysis and methylation of the cod liver oil sample, gives the fat recovery. For SFE, the initial conditions applied were: sample size, 100 mg; supercritical fluid, pure CO_2; extraction at 60°C during 2 min static and 15 min dynamic; flow 1 ml/min; nozzle temperature, 45°C; collection on an ODS trap at 45°C and rinsing with 1.5 ml n-hexane.

10.4.10.3 Results and discussion

The first parameter varied was the density. The results are presented in Figure 10.23.

At 0.4 g/ml, the recovery both for the PCBs and the fat is low; at 0.6 g/ml, 75% of the PCBs is extracted and only 2% of the fat is co-extracted; and at 0.8 g/ml, 88% of the PCBs is extracted but also 72% of the fat. This plot shows the influence of density on the recovery and the difference in solubility of the two compound classes. This difference can form the basis for further optimisation of the selectivity in SFE for that particular case.

From Figure 10.23, densities of 0.6 and 0.5 g/ml were selected as starting points for further optimization, because at these densities relatively high recoveries of PCBs were obtained, while the fat recoveries were low. The data obtained are summarized in Table 10.13 and the following conclusions can be drawn.

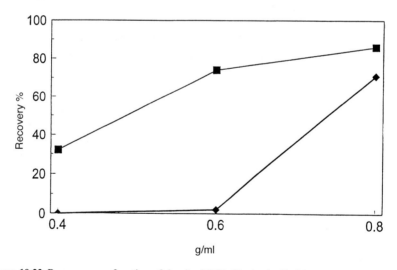

Figure 10.23 Recovery as a function of density (60°C, 15 min, 1 ml/min). Abbreviations: PCBs, polychlorinated biphenyls (—■—); FAMEs, fatty acid methyl esters (—◆—).

An increase of the extraction time from 15 to 30 min at constant density (0.6 g/ml), extraction temperature (60°C) and flow (1 ml/min), resulted in a higher recovery for the PCBs (83%), but also in an increase

Table 10.13 SFE of PCBs from cod liver oil: influence of SFE conditions on recoveries of PCBs and FAMEs

No.	Density g/ml	Temp °C	Time min	Flow ml/min	Recovery % PCBs	Recovery % FAMEs
1	0.4	60	15	1	32	0
2	0.6	60	15	1	75	2
3	0.8	60	15	1	88	72
4	0.6	60	30	1	83	7
5	0.6	60	15	2	104	10
6	0.5	60	15	2	98	2
7	0.5	80	30	1	(140)	15

Abbreviations: PCBs, polychlorinated biphenyls; SFE, supercritical fluid extraction; FAMEs, fatty acid methyl esters.

of fat recovery (7%). An increase of the flow from 1 to 2 ml/min at 0.6 g/ml, extraction time (15 min) and extraction temperature (60°C), resulted in 104% PCBs and 10% fat. From these data, it was clear that flow is an important parameter. Optimized conditions could then be found using a reduced density of 0.5 g/ml at 60°C and an extraction time of 15 min. Under these conditions, the influence of flow and temperature on the recovery were evaluated. The results for flow are presented in Figure 10.24.

At a flow of 2 ml/min, the recovery of PCBs was 98%, while only 2% of the fat was extracted. An increase of the extraction temperature from 60

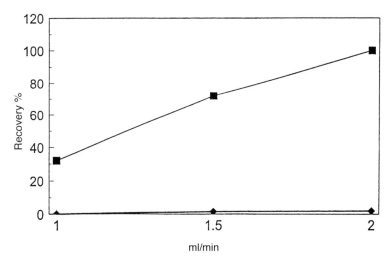

Figure 10.24 Recovery as a function of flow rate (60°C, 15 min, 0.5 g/ml). Abbreviations: PCBs, polychlorinated biphenyls (—■—); FAMEs, fatty acid methyl esters (—♦—).

to 80°C at 0.5 g/ml, a flow of 1 ml/min and 30 min extraction time yielded too high a recovery for the PCBs and an intolerable increase of fat recovery (15%). The recovery of the PCBs could not be measured accurately because of contamination of the CGC-ECD system. In this case, increasing the extraction temperature is not the best choice for increasing the recovery because the selectivity of the extraction is lost. With SFE conditions of 0.5 g/ml density, a temperature of 60°C, extraction during 15 min and a flow of 2 ml/min, a reliable procedure was developed. Repeatablity was good (David et al., 1992). Similar results were reported by Bowadt and co-workers (1994) for the extraction of low concentrations of PCBs from fish tissues.

10.4.11 Supercritical fluid extraction of pesticides in orange juices

10.4.11.1 Introduction

The application selected demonstrates, firstly, the possibilities of SFE as a method of selective sample preparation and, secondly, that the technique can also be applied to the enrichment of trace analytes from aqueous samples. An orange juice sample spiked by an Italian food laboratory with the pesticides thionazin, methyl parathion, fenthion, methidation, pyrazophos, phosalone, vinclozolin and procymidone was sent to the laboratory to evaluate the possibilities of bridged SFE-CGC-MS, compared with the conventional method of LLE followed by clean-up over silica-Florisil or by size-exclusion chromatography. The concentrations of the pesticides were communicated only after submission of the data.

10.4.11.2 Experimental

Initial attempts to extract the aqueous solution as such failed and mixing with an adsorbent was essential. A 2 g amount of Chromosorb W (60–80 mesh) (Alltech) was placed in a 7 ml extraction thimble and 2 ml of orange juice was added on top of the adsorbent, followed by another 0.5 g of Chromosorb W. SFE was performed with a 1 ml/min flow of pure CO_2, at a density of 0.75 g/ml and an extraction temperature of 50°C, during 5 min of static and 30 min of dynamic operation. The extracted solutes were collected on an ODS trap maintained at 20°C. The nozzle temperature was 45°C. After completion of the extraction, the trap was rinsed with 1 ml of chloroform and the vial containing the extract was automatically transferred via the vial transport arm of the HP 7673B autosampler to the GC-MS system operated in the full-scan mode. A 5 μl volume was injected in the splitless mode (45 s), using an increased inlet pressure at the time of injection (pressure pulse 300 kPa for 1 min) on a

30 m × 0.25 mm ID HP-5 MS column with a 0.25 μm film thickness. The temperature of the column was programmed from 50°C (1 min) to 150°C at 20°C/min and then to 260°C at 6°C/min. Helium was the carrier gas at 50 kPa in the constant-pressure mode.

10.4.11.3 Results and discussion

Under the moderate extraction conditions applied, a relatively clean chromatogram was obtained, with palmitic and oleic acid as the main peaks. The selectivity of SFE was fully exploited. The pesticides were soluble at the 0.75 g/ml density of neat CO_2 and an extraction temperature of 50°C. Whereas, both the adsorbent (silica) in the thimble and the collection (ODS) trap helped to retain unwanted matrix solutes, polar compounds, such as the lower free fatty acids, and high molecular mass solutes, i.e. triglycerides, which could interfere (as such or by decomposition products formed in spitless injection) in the elucidation and dosage of the pesticides. The spiked pesticides could be identified in the elution window from 16 to 24 min by comparing the spectra with those listed in a pesticide spectral library. For quantification, the external standard method was applied, selecting a typical ion for each pesticide. Figure 10.25 shows the extracted ion chromatograms for vinclozolin (m/e 285) and procymidone (m/e 283) in the fruit juice sample.

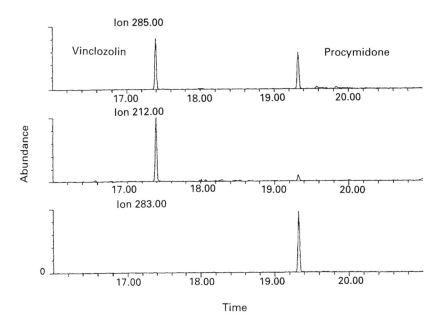

Figure 10.25 Extracted chromatograms for vinclozolin (peak 1, m/e 285) and procymidone (peak 2, m/e 283) in supercritical fluid extraction (SFE) extract of spiked orange juice.

The data are summarized in Table 10.14, giving the quantification ion, the measured concentration (mean value of three experiments), the RSD, the amount spiked in the sample and the recovery.

Table 10.14 Quantification data for pesticides spiked in orange juice and determined by SFE-CGC-MS

Pesticide	Quant. ion	Measured concn. ppb	RSD %*	Spiked ppb	Recovery %
Thionazin	97	17	3.5	20	85
Methyl parathion	263	107	2.9	101	103
Fenthion	168	92	4.1	103	89
Methidation	145	44	3.6	51	86
Pyrazophos	373	28	4.8	30	93
Phosalone	182	429	3.3	505	85
Vinclozolin	285	273	3.7	303	90
Procymidone	283	130	4.1	130	100

*: n = 3. Abbreviations: SFE, supercritical fluid extraction; CGC, capillary gas chromatography; MS, mass spectrometry; RSD, relative standard deviation; concn, concentration; Quant., quantification.

Compared with data obtained by conventional methods of sample preparation, SFE proved to be an excellent technique to quantify the dose of pesticides in fruit juices. Recoveries ranged 85–100% with precisions lower than 5% for the total analytical procedure. Moreover, capillary gas chromatography–mass spectrometry analysis in the selected-ion monitoring mode should allow determinations at the low ppb level. The total analysis time is approximately 1 h and both sample preparation and analysis are fully automated.

References

Anton, K. and Berger, C. (eds.), (1997) *Supercritical Fluid Chromatography with Packed Columns*, Marcel Dekker, New York.
Arthur, C.L., Pratt, K., Motlagh, S., Pawliszyn, J. and Belardi, R.P. (1992) *J. High Resolut. Chromatogr.*, **15** 741.
Ashby, J. and Craig, P.J. (1991) *Appl. Organomet. Chem.*, **351** 173.
Bao, M.L., Pantani, F., Barbieri, K., Burrini, D. and Griffini, O. (1996) *Chromatographia*, **42** 227.
Baltussen, E., Janssen, H.-G., Sandra, P. and Cramers, C.A. (1997) *J. High Resolut. Chromatogr.*, **20** 385.
Baltussen, E., David, F., Sandra, P., Janssen, H.-G. and Cramers, C.A. (1998a) *J. Chromatogr.*, **805** 237.
Baltussen, E., David, F., Sandra, P., Janssen, H.-G. and Cramers, C.A. (1998b) *American Laboratory*, in press.

Baltussen, E., David, F., Sandra, P., Janssen, H.-G. and Cramers, C.A. (1998c) Proc. 20th International Symposium on Capillary Chromatography, CD-ROM, K.01, Publ. IOPMS, Kortrijk, Belgium.

Batussen, E., David, F., Sandra, P., Janssen, H.-G. and Cramers, C.A. (1998d) *Chromatographia*, submitted.

Baltussen, E., David, F., Sandra, P., Janssen, H.-G. and Cramers, C.A. (1998e) Proc. 20th International Symposium on Capillary Chromatography, CD-ROM, M.01, Publ. IOPMS, Kortrijk, Belgium.

Baltussen, E., Snijders, H., Janssen, H.-G., Sandra, P. and Cramers, C.A. (1998f) *J. Chromatogr.*, **802** 285.

Bicchi, C., D'Amato, A., David, F. and Sandra, P. (1987) *Flavour and Fragrance J.*, **2** 49.

Bicchi, C., D'Amato, A., David, F. and Sandra, P. (1988) *Flavour and Fragrance J.*, **3** 143.

Bicchi, C., D'Amato, A., David, F. and Sandra, P. (1989) *J. High Resolut. Chromatogr.*, **12** 316.

Bowadt, S., Johansson, B., Freukilde, P., Hansen, M., Zille, D., Larsen, B. and de Boer, J. (1994) *J. Chromatogr.*, **645** 189.

Buchholz, K.D. and Pawliszyn, J. (1993) *Environ. Sci. Technol.*, **27** 2844.

Buchholz, K.D. and Pawliszyn, J. (1994) *Anal. Chem.*, **66** 160.

Burger, B.V. and Munro, Z.M. (1986) *J. Chromatogr.*, **370** 449.

Ceulemans, M. and Adams, F.C. (1996) *J. Anal. At. Spectrom.*, **11** 201.

Craig, P.J. (1986) *Organometallic Compounds in the Environment: Principles and Reactions*, Harlow, Essex, UK.

Crespin, M.A., Ballastreros, E., Callego, M. and Valcarcel, M. (1997) *J. Chomatogr.*, **757** 165.

David, F., Verschuere, M. and Sandra, P. (1992) *Fres. J. Anal. Chem.*, **344** 479.

David, F., Soniassy, R., Verschuere, M. and Sandra, P. (1993) *Int. Labmate*, Vol. XVIII, Issue VI, 58.

David, F., Sandra, P. and Stafford, S. (1996) Hewlett Packard, Application Note 228-361.

David, F., Sandra, P., Bremer, D., Bremer, R., Roglis, F. and Hoffmann, A. (1997) *Labor Praxis*, **21** (5) 33.

David, F., Correa, R. and Sandra, P. (1998a) Proc. 20th International Symposium on Capillary Chromatography, CD-ROM, P. 52, Publ. IOPMS, Kortrijk, Belgium.

David, F. Baltsussen, E. and Sandra, P. (1998b) *J. Chromatogr.*, submitted.

De Smaele, T., Verrept, P., Moens, L. and Dams, R. (1995) *Spectrochim. Acta Part B*, **50** 1409.

De Smaele, T., Moens, L., Dams, R. and Sandra, P. (1996a) *Fres. J. Anal. Chem.*, **354** 778.

De Smaele, T., Moens, L., Dams, R. and Sandra, P. (1996b) *LC-GC International*, **9** 138.

De Smaele, T., Moens, L., Dams, R., Sandra, P., Van der Eycken, J. and Vandyck, J. (1998) *J. Chromatogr.*, **793** 99.

Dionex, Application Note 316, 1996.

Gerstel Aktuell 1997, p. 4.

Gorecki, T., Mindrup, R. and Pawliszyn, J. (1996) Proc. 18th International Symposium on Capillary Chromatography (ed. P. Sandra), Huthig Verlag, Heidelberg, p. 163.

Grob, K. and Zurcher, F. (1976) *J. Chromatogr.*, **117** 285.

Guiochon, G. (1980) *Anal. Chem.*, **52** 2002.

Haghebaert, K., David, F. and Sandra, P. (1996) Proc. 18th International Symposium on Capillary Chromatography (ed. P. Sandra), Huthig Verlag, Heidelberg, p. 746.

Hawthorne, B., Miller, D.J., Nivens, D. and White, D.C. (1992) *Anal. Chem.*, **64** 405.

Helmig, D. and Greenberg, J.P. (1994) *J. Chromatogr.*, **677** 123.

Hill, S.J., Bloxham, M.J. and Worsfold, P.J. (1993) *J. Anal. At. Spectrom.*, **8** 499.

Jüttner, F. (1988) *J. Chromatogr.*, **442** 157.

Kennedy, G.J. and Knox, J.H. (1972) *J. Chromatogr. Sci.*, **10** 549.

Knobloch, T. and Engewald, W. (1995) *J. High Resolut. Chromatogr.*, **18** 635.

Kopecni, M.M., Tarana, M.V., Cupic, S.D. and Comor, J.J. (1989) *J. Chromatogr.*, **462** 392.
Lesnik, B. and Fordham, O. (1994/1995) *Environ. Lab.*, **Dec/Jan** 27.
Lewis, J.B. (1995) *J. Appl. Chem. Lond.*, **5** 228.
Lobinski, R. (1997) *Appl. Spectrosc.*, **7** 262A.
Lövkvist, P. and Jönsson, J.Å. (1987) *Anal. Chem.*, **59** 818.
Medvedovici, A., Kot, A., David, F. and Sandra, P. (1997) in *Supercritical Fluid Chromatography with Packed Columns* (ed. K. Anton and C. Berger), Marcel Dekker, New York, USA.
Moens, L., De Smaele, T., Dams, R., Van Den Broeck, P. and Sandra, P. (1996) *Anal. Chem.*, **15** 1604.
Nilsson, T., Ferrari, R. and Fachetti, S. (1996) Proc. 18th International Symposium on Capillary Chromatography (ed. P. Sandra), Huthig Verlag, Heidelberg, p. 618.
Noble, A. (1993) *J. Chromatogr.*, **642** 3.
Noij, T.H.M. and van der Kooi, M.M.E. (1995) *J. High Resolut. Chromatogr.*, **18** 535.
Ortner, E.K. and Rohwer, E.R. (1996) *J. High. Resolut. Chromatogr.*, **19** 339.
Pawliszyn, J. (1997) *Solid Phase Microextraction: Theory and Practice*. Wiley, New York.
Peters, R.J.B. and Renesse van Duivenbode, J.A.D.V. (1994) *Atoms. Environ.*, **28** (15) 2413.
Pocurrull, E., Calull, M., Marce, R.M. and Borrull, F. (1994) *Chromatographia*, **38** 579.
Potter, D.W. and Pawliszyn, J. (1994) *Environ. Sci. Technol.*, **28** (2) 298.
Roeraade, J. and Blomberg, S. (1986) *J. High. Resolut. Chromatogr.*, **12** 138.
Rudzinski, W., Gierak, A., Leboda, R. and Dabrowski, A. (1995) *Fres. J. Anal. Chem.*, **352** 667.
Sadtler Standard Gas Chromatography Retention Index Library (1985) Sadtler Research Laboratories, Philadelphia, USA.
Sandra, P., Kot, A., Medvedovici, A. and David, F. (1995a) *J. Chromatogr.*, **703** 467.
Sandra, P., Haghebaert, K. and David, F. (1997) *Int. Env. Tech.*, **6** (5) 6.
Sandra, P., Haghebaert, K. and David, F. (1996a) *Int. Env. Tech.*, **6** (2) 13.
Sandra, P., Mededovici, A., Kot, A. and David, F. (1995b) *Int. Env. Tech.*, **5** (3) 10.
Sandra, P., Medvedovici, A., Kot, A., Vilas Boas, L. and David, F. (1996b) *LC-GC International*, **9** 540.
Smits, R. (1994) *LC-GC International*, **7** 694.
Soniassy, R., Sandra, P. and Schlett, C. (1994) *Water Analysis*. Hewlett-Packard, Publication (23) 5962-6216E, 141.
Sprunar, J., Schmitt, V.O., Lobinski, R. and Monod, J.-L. (1996) *J. Anal. At. Spectrom.*, **11** 193.
Staniewski, J. and Rijks, J.A. (1993) *J. High Resolut. Chromatogr.*, **16** 182.
Tang, Y.-Z., Tran, Q. and Fellin, P. (1993) *Anal. Chem.*, **65** 1932.
Temmerman, I., David, F., Sandra, P. and Soniassy, R. (1991) Hewlett Packard, Application Note 228-135.
Tesarova, E. and Pacarova, V. (1983) *Chromatographia*, **717** 269.
Ventura, K., Príhoda, P. and Churácek, J. (1995) *J. Chromatogr.*, **710** 167.
Verschueren, K. (1996) *Handbook of Environmental Data on Organic Compounds*, Reinhold, New York, USA.
Verzele, M., Dewaele, C. and De Weerdt, M. (1988) *LC-GC*, **6** 966.
Vreuls, R., de Jong, G.J., Ghijsen, T. and Brinkman, U.Th. (1994) *J. Assoc. Anal. Chem.*, **77** 306.
Wilke, C.R. and Chang, P. (1995) *AIChE J.*, **1** 264.
World Health Organization (1984) Guidelines for Drinking Water Quality, Vol. 1. Recommendations. WHO, Geneva.
Wormann, H. and Hoffmann, A. (1994) Proc. 16th International Symposium on Capillary Chromatography (ed. P. Sandra), Huthig Verlag, Heidelberg, p. 416.

Index

absorptive extraction 244
Accelerated Solvent Extraction 139, 146, 223, 233, 239, 240, 253
acetaldehyde 6, 32
acid gases 6
acids 63
adsorptive extraction 244
affinity mode 64
agitation 84, 85, 92
air monitoring 41, 131, 158, 256, 259
aldehydes 244
alkaloids 163
amines 63
amino acids 8
amphetamines 212
anabolic steroids 205
analysis cycle 1
analyte leaching 207
aniline 47
animal feedstuffs 182
annular denuders 30
antibiotics 68
antibody columns 206
antioxidants 68, 135, 224, 232, 238
anti-schizophrenic agent 163
arochlor 158
aromatic hydrocarbons 31
aromatic oils 159
atomic pressure ionisation 255
atrazine 179
automated dynamic sorptive extraction 271
automated sequence trace enrichment of dialysates (ASTED) 208
automated Soxhlet 157, 158
automated SPE 205

backwashing 23
bases 63
beer 70
benzidines 68
beverage testing 40
BHT 226
bioassays 40
biological fluids 40, 41, 195, 196, 202

bleaching pulp 9
blood analysis 212
BTEX 94, 158, 160, 244
bubbler/impinger system 18, 20

capillary electrophoresis 72
carbonyl compounds 7, 68
carboxylic acids 70
cartridge technologies 61
cell disruption 197
cellular membranes 33
centrifugation 214
cereals 70
cheese 162
chimassorb 225, 238
chlorinated acids 68
chlorinated benzenes 183
chlorinated pesticides 68, 160, 162
chlorinated polyisoprene 231
chlorodane 132
chloropyrifos 132
chromatographic mode sequencing (CMS) 68
cigarette smoke 6
citrus oils 47
clay 157
column switching 208
columns, diposable 64
conditioning solvent 57, 64
co-solvents 107
countercurrent extraction 16, 21, 22
Craig extraction apparatus 21
creams 25
cross current extraction 17, 22
crossflow filtration 46, 47, 49
crude oils 8
cylindrical denuders 31

dairy products 46
denaturation 198
dental materials 70
denuders 25, 27, 28, 30, 31
derivatisation 210
designer phases 70
desisopropylatrazine 281

306 INDEX

dialysis 35, 197, 208
dibutyltin dichloride 253
dielectric polarisation 167
diesel particulates 158
dioxins 68, 159, 160
dipolar polarisation 167
diquat 68
direct extraction 79
disk based systems 203, 279
dispersion of sample 151
dissolution 221
distribution ratio 11
distribution/partition constants 10, 82, 95, 129
drinking water 8
drug and drug metabolites 134, 161, 211
drying of samples 151
dust 159
dynamic sorptive extraction 245, 249, 256, 259, 268, 274

electrodialysis 35
electronic polarisation 167
emulsion liquid membranes 49
environmental applications 130, 157, 172
ethylene-vinyl polymer (eva) 225
evaporation 23, 214
exhaustive extractions 20
explosives 161
extraction cell 113, 143
extraction disks 56, 61, 278, 279
extraction efficiency 94
extraction parameters 83, 152
extraction with membrane protection 79

fats 122, 134, 162, 163
fatty acid methyl esters (FAMES) 297
fatty acids 114
fenbofen and felbina 202
fermentation broths 8
filters 28, 29, 30, 32, 33, 40, 197
flow restrictors 115
fly ash 159, 160
food and flavours 70, 133, 162
forensic area 68
formaldehyde 212
freeze drying 197
freons 160
fuels 70
fungicides 189
fused silica fibres 76

gas phase extraction 92, 247, 250, 286
gas phase microwave-assisted extraction 190
gases 7, 35
GC-MS 279
gel extraction 39, 44
glucoronidase and glucoronides 208
grains 160, 162
graphitised carbon black 268
gum phase extraction (GPE) 245, 249

headspace analysis/extraction 79, 80, 93, 190, 212
herbicides 157, 179, 189
high density polyethylene (HDPE) 159, 225, 238
hollow fibre membranes 47, 87, 88
homogenisation 207
hot ball model 226, 228
humic compounds 8, 134
hydraulic fluids 41, 161
hydrazones 2
hydrocarbons 6, 25
hydrolysis 207
hydroxyls 63
hyperfiltration 34, 9, 43

Ibuprofen 210
immuno-extraction procedures 205, 217
impingers 19, 20
impregnated filters 28, 29, 30, 32, 244
insecticides 189
ion exchange 56, 62, 63, 204
ion pairing 109, 201
Irgafos 229
Irganox 159, 226, 234, 235, 238, 239
isocyanates 6, 32, 33

kinetics 84, 106, 119, 125

lactide co-glycoside polymers 136
large volume injection 279
layered phases 62
leaching 5
light stabilisers 135
liquid /liquid extraction 199, 210, 214, 215, 243, 268, 271, 272, 293
liquid /solid extraction 222, 224
liquid membranes 48
liquids 132
lubricants 70

macrofiltration 33, 35

INDEX 307

marine sediment 178
marine tissue 160
matrix solid phase dispersion 205
mayonnaise 162
membrane filters 61
membrane distillation 35
membrane extraction with sorbent interface(MESI) 76, 86, 90, 93
membrane extraction 33
membrane fouling 46
membrane hyperfiltration 43
membrane introduction mass spectrometry(MIMS) 86
method development 66, 118, 138
method validation 126, 154
micro liquid–liquid extraction (µLLE) 248, 253, 261
microwave assisted solid phase extraction 189, 223, 237, 239
microwave assisted solvent extraction 167, 237, 239, 240
miniaturisation 71
mixed mode phases 62
modifiers 107, 108, 123, 136
molecular imprinted polymers (MIPs) 205, 206, 207, 217
molecular imprinting 70
multiple step extraction 16
muscle 8
multiple layer filling 114

nanofiltration 34, 35, 37, 39, 43, 46
naphthanic oils 159
naphthol 8
nicotine 6, 7, 32
nitrogen dioxide 7
nitrophenol 7
nonylphenol 180
normal phase 62, 63
nylon 137, 159, 236

oils 8
ointments 68
optimisation 93, 120, 126
organic colloids 9
organic mercury 25
organochlorine pesticides 158-60
organometallics 288, 290, 293
organophosphorus pesticides 157, 158, 160, 189
oxygenates 9

PAHs 6, 7, 15, 24, 25, 107, 122, 158, 160, 174-78, 182, 184, 189, 195, 244, 256, 286
parathion 209
partition isotherm 13
PCBs 122, 124, 158, 160, 178, 183, 184, 187-89, 195, 244, 286, 296-99
PDMS 89, 94, 245, 255, 271-74, 276
peanut butter 163
peptides 68
peroxyacetyl nitrate 32
pervaporation 35, 45
pesticides 68, 114, 157, 162, 175, 184, 195, 285, 300
petroleum products, 7
Phellodendri cortex 109, 111
phenols 8, 24, 68, 131, 180, 184, 187
phenoxy-acid herbicides 68
phthalate esters 189, 190
plasma 68, 209, 211, 215
plasticisers 135, 236
polyaromatic carbons 25
polybutyleneterephthlate (PBT) 137, 159, 234
polychlorinated dibenzofurans (PCDD/F) 33
polyethylene 136, 159, 222
polyethylenetetraphthlate 41, 136, 137, 222
polyisocyanates 32
polymer additives 124, 161, 162
polymeric materials 135
polyolefins 222, 224
polyorganophosphazenes 48
polypropylene 159, 222, 226, 238
polystyrene 136, 159
polyurethane foam 177
polysulphone 48
polyurethane 131
polyvinylchloride 41, 136, 159, 224
pressurised fluid extraction (PFE) 146
programmable temperature vaporisation (PTV) 254, 279, 280
protein precipitation 198

reprecipitation, 221
restricted access media 209
retention data 129
reverse osmosis 35
reverse phase 63
robotics 58, 59, 201, 213, 214, 219

sample application 65
sample morphology 121
sample size 121

sediment 159, 160, 176, 177, 180-82, 250
sensitiser 168
silicone membrane 88
size exclusion 64, 116
sludge 250
soils 133, 152, 157, 159-61, 174-78, 182, 188, 190, 250, 286, 288
solubility data 129
solute collection 116
solvent effects 168
solid phase trapping 117
sonication 101, 132, 225
Soxhlet extraction 19-21, 103, 128, 132, 137, 138, 146, 159, 163, 197, 223-25
Soxtec 225
Soxwave 169
solid phase extraction (SPE) 54, 115, 201, 202, 204, 205, 207, 214, 215, 278, 248, 268, 278, 279, 281, 283-86
SPE extraction plate 71
SPE pipette 71
solid phase micro extraction (SPME) 75, 212, 245, 249, 266, 268, 271-74, 290-93, 295, 296
static and dynamic sorptive extraction 246, 249, 253, 268, 288

static extraction mode 113
styrene butadiene rubber 159
sulphonated azo dyes 68
supercritical fluid chromatography 108, 111, 115, 129, 256, 285
supercritical fluid extraction 101, 223, 226, 234, 239, 251, 253, 283, 284, 297, 299-302
supercritical fluids 102
supported liquid membranes 87

thermodynamics 81
tinuvin 224, 225
total petroleum hydrocarbons (TPH) 153
toxicokinetics 195
tris-nonylphenol phosphate (TNPP) 225

urine 215

viscous trapping fluids 117
volatile organic compounds (VOCs) 93, 183
volume reduction 209

water analysis 183, 184, 261, 266, 268, 271, 274, 278, 281